Introducing Sea Level Change

T0261341

Other titles in this Series

For further details of these and other Dunedin
Earth and Environmental Sciences titles see
www.dunedinacademicpress.co.uk

INTRODUCING
SEA LEVEL CHANGE

Alastair Dawson

EDINBURGH ◆ LONDON

Published by
Dunedin Academic Press Ltd
Hudson House
8 Albany Street
Edinburgh EH1 3QB
Scotland

www.dunedinacademicpress.co.uk

ISBNs
9781780460871 (Paperback)
9781780466002 (ePub)
9781780466019 (Amazon Kindle)

British Library Cataloguing in Publication data
A catalogue record for this book is available from the British Library

Typeset by Makar Publishing Production, Edinburgh
Printed in Poland by Hussar Books

For Sue, Greg, Laura and Euan.
Particular gratitude is expressed to the late Brian Sissons
who inspired me to study sea level change.

Contents

Preface

One of the most important issues in current debates on climate change is sea level rise. Not a day goes by when there is not a newspaper article, TV or radio presentation on the topic. With millions of people around the world living at the coast there is real concern that a future dramatic rise in sea level will cause chaos to coastal towns and cities. Yet, the world has experienced many dramatic changes of relative sea level in the past. In this book I give an account of the remarkable history of these past changes. It is a story that takes us from the last ice age on Earth to the effects of present-day climate changes on the world's glaciers, ice sheets and oceans. I describe the rates and patterns of sea level change that have occurred over the last century and show how scientific advances are being used to measure present-day patterns of relative sea level change. This book provides the reader with the key scientific information that is needed om which to base an informed discussion on the topic of 'sea level rise'. It is hoped that it will be of value, not only to students as an educational textbook, but also to a wide range of non-governmental organisations as well as to the general public wishing to understand the subject.

Alastair Dawson
Dundee, October 2018

Note: all terms initially highlighted in **bold** are defined in the Glossary at the end of the book.

List of illustrations and tables

Unless otherwise acknowledged in the captions all illustrations are the work of the author.

1   What does 'sea level' mean?

Introduction

To many, the issues of sea level change and sea level rise, together with the dangers that society may face in the future, are very simple – global warming caused by climate change is melting glaciers, and this is causing sea level to rise uniformly worldwide. This view, as we shall see, is light years away from reality. Although based on good intentions to save our planet from the effects of climate change, this interpretation of sea level change is at best misinformed and at worst just plain wrong. This book is essentially a rough guide to the science of sea level change. It attempts to provide a relatively simple and straightforward explanation of an otherwise highly complicated topic. So why is the science described in the media and even in some academic publications misinformed? The problem arises because sea level change is a difficult thing to measure.

The view from the beach

Imagine being able to live for a thousand years and every morning walking down to the beach. Over time you notice that the sea seems to be encroaching on the land. The conclusion is that sea level is rising and this, in turn, is leading to increased coastal erosion and flooding. However, viewed from the beach, that statement may not be true. For example, suppose the sea had remained at the same position for several hundred years and the land had been sinking. Viewed from the beach, it would appear that sea level had been rising over time, but in fact this interpretation would be wrong (Fig. 1.1).

Suppose instead that over these same several hundred years the sea had indeed been rising, let's say by 3mm/yr over the course of 200 years, and has resulted in 60cm of rise. But imagine that the land had been rising also at the same rate and by the same amount – the view from the beach would be that no change had taken place. Another scenario might be that over 200 years the sea had fallen by, say, 40cm. But if the land had subsided by 60cm over the same period, the view from the beach would be that sea level had risen by 20cm. The view from the beach, therefore, can only tell us about changes in relative sea level. It cannot distinguish for us absolute vertical changes in the sea or the land.

What this tells us is that the view from the beach tells us more or less nothing about sea level change – all it can do is inform us about relative

Figure 1.1 The Barbados view of relative sea level from the beach. Image credit: Shutterstock/Filip Fuxa.

changes in sea level. But since all of us view the ocean from our respective beaches, we have no direct experience of the phenomenon of global sea level change except our view of relative change from our own beach. How then do we proceed? Fortunately we can turn to scientists to explain to us the view from space.

The view from space

Viewed from space, the Earth's sea surface is a remarkable sight (Fig. 1.2). First and foremost it is not flat. In fact it is nowhere close to being flat. Instead it is characterised by a surface topography that exhibits huge differences in elevation between one ocean and another and also across individual oceans. Viewed from space, parts of the North Atlantic Ocean bulge up in places by as much as *c*.60m above average. By contrast, in the middle of the Indian Ocean the sea surface descends to levels close to –110m below the average. Viewed from a cruise ship sailing across the Indian Ocean, the sea surface appears flat; the horizon is indeed where it should be. But viewed from space our cruise liner is sailing downslope as it approaches the 'low' area of the Indian Ocean and then sailing upslope to leave this area. Why should sea level be so low in this area? A popular view is that it is caused by a deficiency of mass within the Earth's mantle beneath the northern Indian Ocean at depths between *c*.300–900km.

The reason that the view from space is so strange is because ocean water, like every other mass on Earth, responds to gravity. In some ocean areas the gravitational pull towards the centre of the Earth is high while in other areas it is low. For the world's oceans this results in a gravitational equipotential ocean surface, known as the geoidal sea surface. Not only that, but large masses of ice on Earth also exert a gravitational pull on surrounding ocean water. This means that in respect of the Greenland and Antarctic ice sheets the ocean surface increases its elevation by several tens of metres as one approaches each ice sheet. Viewed from space this gravitational pull decreases in strength at progressively greater distances from the respective ice sheet. But from space, the effect of this pull can be measured even at distances of *c*.1000–1500km away from the ice edge. Viewed from a ship this phenomenon is invisible, but viewed from space it is very clear.

The view of the scientist

The notion of 'sea level change' means different things to different groups of scientists. To the geologist who observes coastal sediments located well above the present shoreline, it prompts the notion that, at a given point on the

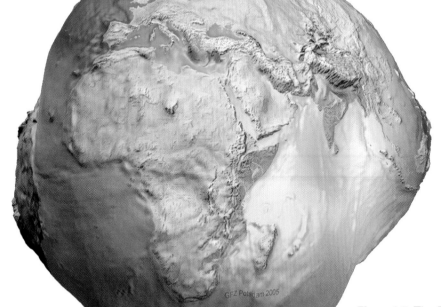

Figure 1.2 The Potzdam gravity 'potato' showing the Earth's geoidal equipotential surface with a greatly exaggerated vertical scale. Image credit: F. Bathelmes and W. Kohler (Journal of Geodesy, 2016, 90).

Earth's surface, sea level must have been higher than present at some time in the past. By attempting to estimate the age of a particular coastal deposit and measuring its height above sea level, the scientist can begin to make inferences regarding the past pattern of relative sea level change at a particular location. By measuring thousands of such samples at different locations, groups of scientists can reconstruct for a given area past patterns of relative sea level change. For geologists interested in climate change, the reconstruction of past ice ages and the melting of past ice sheets, such information is invaluable. For example, as the ice sheet melted at the close of the last ice age, the sea could penetrate into areas gradually being uncovered by the melting ice. In this way, study of ancient marine sediments laid down at this time can tell a great deal in respect of the patterns of ice sheet melting that took place. In some coastal areas traces of former sea levels lie submerged beneath the present sea surface. Evidence of such drowned landscapes include, for example, submerged archaeology testifying to patterns of human settlement related to former changes in relative sea level.

For scientists working in warmer climates, the reconstruction of past changes in relative sea level is just as important. For archaeologists, periods of lowered sea level in the past might mean land bridges and important routes of past human migration. For ecologists it might contribute to understanding the long-term evolution of plants and animals. But the issue of future sea level rise is fundamental to future life on planet Earth. Since a large proportion of the world's population lives close to or near the coast, it is vital that we understand the processes responsible for sea level change. For our modern world we have to be able to measure modern-day rates of change with a high degree of accuracy. Just as important, we need to be able to predict as best we can how sea level is likely to change in the future. A cornerstone of such studies is to understand past changes in relative sea level for different areas of the world. A fundamental tenet in geology is that an understanding of the past is key to understanding the present. As a result, great efforts have been made by scientists around the world over the last century, trying to reconstruct past patterns of relative sea level change – from icy northern latitudes to isolated Pacific islands – so that we can better understand present changes and help predict future change.

For geophysicists, such studies provide priceless information about the Earth's structure. Any change in sea level alters the radius of the Earth and can lead to small changes in the Earth's rotation. Equally, the growth and decay of major ice sheets in the past has resulted in major changes in the redistribution of mass (ice and water) across the Earth's surface, which caused not only changes in the Earth's spin rate but also slight changes in the axis of the Earth's rotation, as well as changes in its radius and tidal dynamics.

Inconvenient truths

Some years ago the American politician, Al Gore, brought the issue of sea level rise into the international spotlight. In his film *An Inconvenient Truth* he described an imminent catastrophe in which most of the world's major cities located close to the coast would soon be inundated by the sea. Graphic images were created to illustrate this looming catastrophe. For example, an image of London was shown in which the Houses of Parliament were half-submerged. One did not have to look far to find maps of the world that showed the possible scale of this impending sea level rise catastrophe. The most dramatic map showed what the world would look like if the Antarctic ice sheet was to melt and sea level was to rise by *c*+60 metres and inundate many of the world's largest cities.

The debate on a future catastrophic rise in world sea level was not confined to descriptions of what would happen to the world's largest cities. It became a prominent issue in remote corners of the planet. Notable amongst these were the islands of the Maldives in the Indian Ocean, whose President during the 1990s was warning that his low-lying islands would soon disappear beneath rising seawaters. There were numerous discussions in the media of the need to evacuate those island communities at greatest risk from future flooding. Needless to say the islands are still there!

More recently, during the final months of the US presidency of Barack Obama, the National Oceanic and Atmospheric Administration (NOAA) produced a report in which they described an extreme scenario where sea level could potentially rise by *c*.2.5 metres by AD 2100. Yet a few months later under the presidency of Donald Trump, the sceptical view of climate change came to the fore, with the United States pulling out of the Paris accord on Climate Change and putting forward the view that the world was not in danger from climate change, and certainly not under threat from the prospect of sea level rise. Where then does this leave our thinking on the issue of sea level rise – should we be worried or not? Are we to believe the 'climate sceptics' that the issue of 'sea level rise' is an illusion? Or are we to take seriously those who warn that future sea level rise will soon destroy our world? What are we to believe, and what information do we draw upon to form our opinions?

In this book we consider the key processes that drive sea level change. We need to know what patterns of sea level change have taken place in the past, and we need to understand the causes of these changes. We also need to understand the timescales of past sea level change, both in history and in prehistory and, from this, we need to learn how measured rates of change taking place at present compare with rates of change estimated for the past, and how these processes might change in the future.

There are many ambiguous statements made regarding the issue of sea level change. Not only does one despair about what is written in the media, but on some occasions a lack of understanding finds its way into academic debate. Scientists describe their own versions of how they understand the issues surrounding the sea level change debate, but quite often these versions are misguided. The reason why so many people form mistaken views has its roots in the division of science into many different disciplines. To understand the issue of sea level change well, one has to delve into a wide range of scientific disciplines to find answers explaining a large range of environmental

processes. If one is to understand the subject to a reasonable degree of competence, one has to investigate aspects of the disciplines of geology, physical geography, oceanography, archaeology, climatology, geophysics, geodesy and glaciology. For scientists the challenge of undertaking inter-disciplinary research is a daunting proposition, but it can also be extremely rewarding. We thus start this book with the challenge of making sense of this incredibly interesting and complicated subject. But we must first ask a fundamental question.

What do we mean by 'sea level' and how do we measure it?

As we look out towards the horizon we see an essentially flat sea level surface disturbed by wind and waves, but how do we measure it? Along coastlines bordering the Mediterranean Sea we are used to 'sea level' remaining in an essentially 'fixed' position because the tidal range is negligible. Yet along most coastlines of the world, the sea experiences daily vertical changes due to the tides. How then are we supposed to measure the vertical position of 'sea level' along coastlines where it is constantly changing? In such areas we can measure the position of such elements as low tide, high tide, high water mark of spring tides, highest astronomical tides, etc. If we measure such positions for long enough using tide gauges, we can derive an estimate of the position of 'mean sea level' (Fig. 1.3). Some tide gauge records are very long indeed. For example, tidal measurements have been made for Amsterdam since AD 1700, while Liverpool has a tidal record dating from AD 1768 and tidal records were first made for Stockholm in AD 1774. By

Figure 1.3 Tide gauge, Reykjavik, Iceland. Image credit: Creative Commons – Luc Van Braekel.

themselves, tide gauge records gathered over substantial periods of time provide detailed records of past changes in local sea level. But many such records are compromised by vertical changes in the level of the land. For example, New Orleans is located over an area of land subsidence that has the effect of giving the impression that rates of sea level rise are higher than they actually are (Fig. 1.4). In this area, rates of ground subsidence have increased as a result of groundwater pumping leading to land

Figure 1.4 Rates of land subsidence, Jefferson Parish, New Orleans, June 2009–July 2012 as seen by the NASA UAVSAR instrument. The exceptionally high values are mainly due to sediment compaction coupled with groundwater pumping. Image credit: NASA/JPL-Caltech, Esri.

subsidence, with values of relative sea level rise in the order of 10mmyr^{-1} not uncommon. Such rates of land sinking pale into insignificance when compared to Jakarta. Here, the use of groundwater pumping through the use of enormous numbers of private wells has contributed to local rates of land subsidence reported to be as high as c.250 mmyr^{-1}! By contrast, some tide gauges elsewhere in the world record changes in relative sea level in areas where significant crustal uplift is taking place. Stockholm is a good example of such a location, where long-term rebound of the land surface caused by the melting of the last ice sheet across Scandinavia is sufficiently great to cause a present-day relative lowering of sea level.

How then is it possible to link together and correlate the positions of 'mean sea level' across different areas of the world? Over a century ago, an engineering project convincingly demonstrated how difficult a task it would be to interpret such differences. During the construction of the Panama Canal it was discovered that on the Pacific coastline the difference between high and low tide can be as much as c.6 metres yet on the Atlantic side it is only c.1 metre. Thus the position of 'mean sea level' is different at each end of the Panama Canal, the water on the Pacific side being on average 20cm higher than that on the Atlantic coastline. Although these different values are due to processes associated with the circulation of ocean currents, they highlight how difficult it used to be to link together the positions of 'mean sea level' across different areas of the world.

How have changing sea level positions been measured?

Prior to the advent of satellite technology, the construction of topographic maps always involved decisions by map-makers about how to define and depict the position of sea level. The need to define such a datum level was a necessity for a multitude of practical reasons such as the construction of buildings, canals and drainage systems, etc. For most countries, a reference location at the coast was chosen as a 'datum' point considered representative of mean sea level. Thereafter all topographic surveys of the landscape using theodolites and accurate levelling used this value as the reference datum. In the United Kingdom, for example, the first levelling survey made use of a benchmark at St John's Church, Liverpool in 1840. This was changed a few years later to a tidal marker pole in Victoria Dock, Liverpool. Routine tidal measurements were made on the marker pole to measure vertical tidal changes and also the position of mean sea level. Some years later (1912–21) a second survey levelling exercise was undertaken for the whole country. At this time a new reference location was established at Newlyn, Cornwall and a new set of benchmarks established across the UK landscape. Later, between 1951and 1956 a third geodetic levelling of the

UK was completed using the Newlyn datum. The Newlyn tide gauge record thus provided a time series of tidal changes for over a century. All other tide gauge stations constructed since then on the British mainland provide records of past sea level changes, but all of them make use of mean sea level at Newlyn as the reference level. There remained the difficult problem, encountered also in other countries, of how to connect islands into a national datum. In most such situations, all that could be achieved was to make use of a local datum for each island area.

Similar methods have been used in other countries to establish the position of mean sea level. In the United States, for example, topographic surveys enabled the creation of a Sea-Level Datum in 1929 that was based on measurement of tidal changes at 26 tide gauge stations throughout the United States and Canada. Whereas the early topographic surveys were based around a reference datum at Meades Ranch in Kansas, a more recent North American Datum (NAD) established in 1983 has no reference datum, and instead makes use of satellite data (see below). In Greece, scientists make use of the Hellenic Geodetic Reference System (1987) that is based on measurement of a central pedestal location at the Dionysos Satellite Observatory north-east of Athens.

These examples from the UK, USA and Greece are given here simply to point out that for most of the twentieth century, individual countries made their own individual efforts to establish national datum levels that were then used to define the position of mean sea level. However, it proved extremely difficult to link the different datum levels together, particularly between countries separated by ocean areas. As the establishment of the North American Datum illustrated, this problem disappeared with the advent of satellite technology.

Mapping the geoidal sea surface and understanding gravity

When satellites introduced the possibility of making new maps of the Earth from a different perspective, our concept of what 'sea level' really was changed dramatically. It soon became clear that the Earth's sea surface was by no means flat (Fig. 1.2). The theory that the Earth's sea surface was undulating rather than flat had been known for a long time, but was not demonstrated until satellite measurements showed significant regional variations in the altitude of the global sea surface. The key to understanding these observations of a 'potato-shaped' Earth surface is gravity.

Suppose that it was possible to remove all the tides and ocean currents, storm waves and swell from the oceans. If this was possible, one would view from space a smooth undulating ocean surface. In some areas the ocean would bulge upwards where the pull of gravity is high, in other areas the ocean surface would sink down where the pull of gravity is much lower. The differences in gravitational pull between 'low' and 'high' are due to spatial variations in the distribution of mass beneath the Earth's crust. As sub-crustal material moves within the Earth's mantle, certain areas within the Earth are characterised by plumes of hot and relatively low-density material that rises upwards. Since the lower-density material has relatively less mass, it creates an area of relatively low gravitational pull that, in turn, causes a relative lowering of the ocean surface. The most extensive ocean area of geoidal sea surface lowering is across the northern Indian Ocean, where it descends to c.–110m below the average elevation for the whole of the world's oceans. By contrast, a well-defined geoidal sea surface high up to +60m occurs across the North Atlantic Ocean

This undulating 'potato shape' of the world's oceans is known as the geoidal sea surface. Using complex mathematics, scientists have theoretically extrapolated this surface across individual continents between one ocean and another to create a numerical model of the Earth's gravitational field which, in turn, reproduces a geoidal sea surface that takes into account the regional effects of gravity on the ocean surface (Fig. 1.2).

The mathematical algorithms of this gravitational model are routinely built into GPS receivers for measuring location and altitude. Thus in the case of the geoidal sea surface low of c.–110m in the Indian Ocean, a GPS instrument will not show a ship sailing downhill towards the centre of the Indian Ocean. Instead the GPS is calibrated to zero so that the ship is always sailing across a level ocean surface. But viewed from space, the geoidal sea surface will indeed exhibit a topographic low in the centre of the Indian Ocean.

The use of satellite altimetry to measure directly the altitude of the ocean surface has been routine since c.1992. At this time the launch of the TOPEX/Poseidon satellite marked the first of c.62,000 orbits of the Earth until it closed down in 2006. The main use of the satellite was to map the topography of the world's oceans through the use of radar altimetry. It followed the earlier Seasat satellite and provided repeated measurements of the world's ocean surface to an accuracy of 3.3cm. The TOPEX/Poseidon mission was thereafter followed by the Jason-1, Jason-2 and Jason-3 missions, the latter of these having been launched in January 2016.

More recently the use of satellite radar altimetry has been supplemented by precise measurements of changes in the Earth's gravity field. The year 1992 marked the start of the Gravity Recovery and Climate Experiment (GRACE) whereby measurements were made of short-term changes in the distribution of mass across the Earth and its variation with time (Fig. 1.5). Gravity field changes include measurements across areas where crustal rebound is still continuing as a result of the disappearance of former ice

sheets. They also provide information on changes in pressure exerted on the ocean floor due to changes in the mass of the world's oceans.

The GRACE programme has been supplemented by the Global Sea Level Observing System (GLOSS) that makes use of tide gauge data from several hundred coastal locations to monitor longer-term regional changes in relative sea level and also to calibrate the radar altimeter data.

In recent years, our understanding of the GRACE data has also been advanced by results from another geodetic space mission, the Gravity Field and Steady State Ocean Circulation Explorer (GOCE), which has enabled scientists to separate the Mean Sea Surface (MSS) topography from that of the marine geoidal sea surface. Subtraction of these two surfaces has allowed scientists to recognise and measure the key features of global ocean circulation as well as the effects of weather systems. For example,

hurricanes, typhoons and severe frontal-cyclones in the middle latitudes normally elevate the ocean surface as a result of lowered air pressure.

Ever since the early 1990s, satellite altimeters have provided measurements of sea surface height. For example, a joint project between NASA Jason-1 and the French Space Agency (CNES) enabled radar altimeters to measure precise distances between the spacecraft and the ocean surface. Once the 'noise' of surface waves was removed, an exceptionally detailed map of the geoidal sea surface could be produced. Not only that, but the data has also been used to produce a gravity model of the world's ocean floor.

It should be noted that errors exist in the measurement of the position of the sea surface using radar altimeters. The trajectory of the orbiting satellite is crucially important – for example, whether individual satellite tracks cross over each other, or whether they follow near-identical orbits over time. Furthermore, the radar pulse measured by the satellite can also be affected by ocean waves that scatter energy from their crests and focus energy within individual wave troughs. Wind speed is also a factor, causing the radar return pulse to be diminished as wind speeds increase. The best method of measurement is through the use of radar interferometry where two or more radar images are used from approximately the same position in space to calculate sea surface altitude.

In addition to the effects of waves and tides that need to be removed from the radar altimetry data in order to map the marine geoidal sea surface, it is also necessary to make allowance

Figure 1.5 Artist's impression of twin GRACE (GRACE-FO) satellites. Image credit: NASA/JPL-Caltech, Esri.

for surface ocean currents. One of the most well-known examples is that of the Gulf Stream in the North Atlantic, where changes in the position and strength of the current can routinely create changes in sea surface altitude of over 1m. Similarly, in the Pacific Ocean, the sea surface is capable of both rising and falling in the order of 0.5m during major El Niño events, while at the coast these altitude changes are recorded on tide gauges.

The measurement over time of the surface of the world's oceans using radar altimetry provides us with information of changes over time in the overall volume of the world's oceans. But this volume measurement is not only affected by how much extra ice melt is added (or sometimes subtracted); it is also influenced by changes in the volume caused by increase or decrease in the density of ocean water. For example, increases in sea temperature can cause changes in the density of ocean water (changes in salinity also affect ocean water density). If the density of ocean water is lowered, ocean volume will increase as a result. These latter changes are referred to as steric changes, namely as they involve changes in ocean volume without any change in ocean mass (see also chapter 11). Thus, satellite altimetry measurements provide us with an estimate of ocean volume that includes *both* any change in ocean mass plus any steric effects. By contrast, the GRACE satellite provides information on ocean mass only. Therefore, by subtracting the GRACE data from the altimetric measurements, it is possible to isolate the effect of steric processes.

Measuring steric changes

During the latter part of the twentieth century the steric component of recent changes to the global ocean surface consisted primarily of ocean temperature measurements, and the majority of these were in respect of sampling to water depths shallower than $c.700$m. However, during the last $c.15$ years the ARGO program resulted in the installation of extensive arrays of floats across the surface of the world's oceans. During the last few years ARGO floats have been able to sample down to depths of $c.2000$m and measure vertical profiles of ocean temperature, pressure and salinity (Fig. 1.6). Some ocean areas have experienced water temperature lowering at different depths within the water column. For these areas, the water masses have experienced thermal contraction (decreased ocean volume), which has contributed to a lowering of the ocean surface. However, for most of world's oceans and for the majority of water column within the first $c.1$km water depth there has been an overall increase in ocean temperature in recent

Figure 1.6 An ARGO aquatic robot that measures vertical profiles of ocean temperature, pressure and salinity across the world's oceans. Image credit: Argo Program Germany/Ifremer.

years (increased ocean volume) that has led to an overall thermal expansion of ocean water. Oceanographers repeatedly measure this global thermal expansion component over time, and have attempted to quantify its contribution to sea level rise (see chapter 10). In an ideal world, the measured trend over time (based mostly on ARGO data) ought to be similar to the trend derived from the subtraction of GRACE and satellite altimeter data. Warming of the global oceans is now an extremely important process causing increases in ocean volume and hence sea level (Fig. 1.7).

Measuring long-term sea level changes

The volume of the world's oceans is approximately 1322×10^6km^3 and has an area of 361.9×10^6km^2. It should be noted that 1km^3 of water has a mass of one Gigatonne (1Gt) and that 1km^3 of ice has a mass of 0.9Gt. Since the area of the world's oceans is approximately 362×10^6km^2 it follows that 362Gt of meltwater is required to raised sea level around the world on average by 1mm. It also means that the melting of 398Gt of glacier ice is required to raise average sea level by the same amount. Recent satellite measurements by NASA suggest that the average annual net loss of ice from Greenland (accumulation–ablation) is in the order of 285Gtyr^{-1}. These figures provide a sense of the scale of environmental factors resulting from changes in the extent and volume of ice sheets, ice caps and glaciers that contribute to changes in relative sea level.

The environmental changes that took place during the last ice age are of quite a different scale and magnitude, and they tell us a great deal about the contrasts between a glacial and an

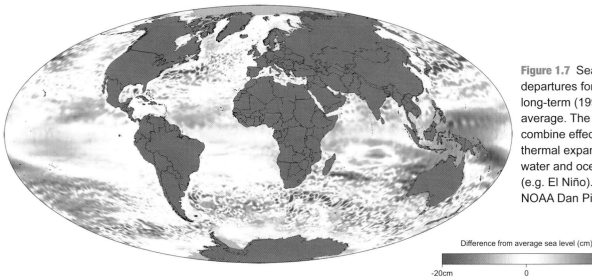

Figure 1.7 Sea level departures for 2011 from long-term (1993–2011) average. The departures combine effects caused by thermal expansion of ocean water and ocean circulation (e.g. El Niño). Image credit: NOAA Dan Pisut.

Difference from average sea level (cm)

-20cm   0   +20cm

interglacial age Earth. For example, during the maximum of the last ice age in Antarctica, had the ice later completely melted, there would have been enough ice to raise average sea level by c.85m. However, it did not do so. Today's Antarctic ice sheet contains a sea level equivalent of c.73m, demonstrating a loss of only 12m global sea level equivalent. By comparison, the present day Greenland ice sheet is much smaller (7.2m of global sea level equivalent) compared to its ice age counterpart (9.8m of global sea level equivalent). On an even smaller scale, the world's c.150,000 glaciers and small ice caps presently contain a sea level equivalent of c.0.3m while their glacial age counterparts amounted to a sea level change equivalent of c.3.9m.

But the biggest changes between the last ice age and the present in terms of glacier ice stored and lost is in respect of the two largest ice sheets that developed in the northern hemisphere during the last ice age. By far the larger of the two was the Laurentide ice sheet in North America, as well as the Cordilleran ice sheet that merged with it across western North America. According to Bill Hay (see list of further reading), during the Last Glacial Maximum (LCM) c.21,000 years ago, the Laurentide ice sheet had a volume of ice equivalent to a sea level equivalent of c.73m, while the Cordilleran ice sheet contains an additional 8.9m of sea level equivalent. The last ice sheet in northern Europe, known as the Fennoscandian ice sheet (or Eurasian ice sheet), extended from the British Isles in the west and as far east as European Russia and central Siberia. It was an enormous ice sheet, but still only half the size of the Laurentide ice sheet with a sea level equivalent of c.42m.

Considered together, and taking into account possible changes in the sizes of the ice sheets in Greenland and Antarctica, Bill Hay concluded that the best estimates of the volumes of ice locked up in the various ice sheets during the last ice age, together with ice stored in the world's glaciers and ice caps, approximated c.228m of sea level equivalent while the ice volume contained within contemporary ice sheets and glaciers around the world is equal to c.83m. The inference, therefore, is that the difference in averaged sea level change equivalent between the last glacial maximum and the present day is c.145m. If these figures are correct, it would imply that the average global sea level fall around the world during the last glacial maximum was in the order of –145m. In the following chapters we will discover to what extent these estimates are correct. We will also learn of the great significance that calculations such as these have for our understanding of the nature of present-day changes in relative sea level and their causes.

2 Evidence for former relative sea level changes

Introduction

Ever since the beginning of the twentieth century when scientists began detailed mapping of relict coastal landforms associated with former sea level positions, progress in unravelling past patterns of change has become more and more sophisticated. In the early days of geological mapping, most attention focused on areas that experienced crustal rebound as a result of the melting of ice sheets. It was in such areas, after all, that the field evidence for past changes in relative sea level was very clear. Through detailed mapping geologists were able to trace ancient shorelines for many kilometres. In some instances, raised shorelines could be traced as far inland as contemporary glacial deposits, thus enabling important links to be made between patterns of relative sea level change and past ice sheet and glacier fluctuations.

In lower latitudes, traces of higher relative sea levels could also be observed. Frequently such evidence took the form of raised beaches and backing cliffs. In other areas, evidence for the occurrence of past sea levels higher than present took the form of elevated coral reefs. Whereas the presence of well-developed raised shoreline features are relatively rare in the landscape, there are many locations where the only evidence for past changes in relative sea level consists of isolated sediment exposures.

In the following pages, a description is given of the different types of erosional and depositional coastal landforms that are routinely used as evidence for past fluctuations in sea level. The list is not comprehensive, with greatest attention being given to the altitudinal relationships between individual coastal landforms and former positions of sea level. The discussion is prefaced by a definition of 'raised beach', a term that is widespread, often misused in the scientific literature. The term 'raised beach' is generally used to refer to any coastal landform that occurs above present sea level, e.g. a shore platform, marine terrace or storm beach ridge, etc. As a result, the term refers to a group of coastal landforms. Thus, descriptions that describe specific coastal landforms as raised beaches (e.g. an uplifted beach ridge) are not helpful. For the scientist, apart from recognising a specific raised coastal landform in the field, the key issue is being able to assign a former position of mean sea level to the landform in question. Only in this way can accurate reconstructions of former patterns of relative sea level change be undertaken (see below).

Erosional shoreline features
Shoreline notches
Shoreline notches present contrasting relationships to sea level (Fig. 2.1). Where an emerged notch consists of

Figure 2.1 Shoreline notch, Fiji, produced by preferential erosion of Holocene emerged reef beneath igneous boulder. Image credit: John Terry.

several rock strata with differing resistances to marine erosion, the morphology of individual notches may reflect a greater influence due to bedrock geology than to any influence of former sea level. Thus many examples exist where the positions of individual notches are more dependent on lithological differences than on the position of former sea level. Some notches are well developed in relatively tideless seas (e.g. the Mediterranean); others occur in areas of high tidal range. Some notches are sufficiently large to represent undercuts in coastal cliffs. Many are produced by process of abrasion and hydraulic impact while others, for example along limestone coastlines, are predominantly formed as a result of bio-erosion. The principal bio-erosional processes responsible for notch development include salt-weathering together with the roles played, for example, by limpets (e.g. *Patella* spp.), and the pelecypod *Lithophaga lithophaga*. The growth and decay of barnacles, mussels, oysters and calcareous algae also contribute to the degradation of coastal rock surfaces. In the case of notches produced by bio-erosion, the inner extent of the concave notch is often interpreted as the position of mean sea level. It should also be noted that limestone and similar rock types with a high calcium carbonate content are soluble in both fresh and salt water, giving rise to karst-type solution features across coastal rock surfaces.

Shore platforms

Shore platforms are relatively flat rock surfaces that dip gently seawards. They are ubiquitous features of many coastal areas and are typically found in association with a backing cliff and occasionally sea caves, notches and rock arches. Relict platforms occur both above and below present sea level. Typically, the platforms may range in width from tens of metres to hundreds of metres and are developed in nearly all rock types. Most platforms exhibit gentle slopes from landward to seaward, although the detailed topography of most shore platform fragments is strongly influenced by local variations in rock lithology. The formation of shore platform remains problematic since, as DW Johnson stated as long ago as 1919, as a platform becomes wider, 'the waves will exhaust themselves on a platform of its own carving'. How wide, therefore, can a 'shore platform' become?

This question is made more difficult to understand since the widest shore platforms across the world are located in high-latitude environments. These features, known as strandflat, range in width from hundreds of metres to several kilometres (Fig. 2.2). In some cases the backing cliffs and cliff-platform junctions are clearly defined, but it is often the case that the platform surfaces are irregular in outline. In most cases where they occur, the rock surfaces are ice-moulded, thus demonstrating that they pre-date at least one period of general glaciation.

In some parts of the world (e.g. Scotland, Norway, Alaska) emerged shore platforms and cliffs originally considered as classic 'wave-cut platforms' are now interpreted as having been produced during periods of former cold climate as a result of frost-shattering on coastal cliffs and removal of debris by sea-ice transport. The most well-known examples are the Main Rock Platform (Main Lateglacial Shoreline) of western Scotland and the Norwegian Main Line, both of which are typically developed in resistant metamorphic

Figure 2.2 Ice-moulded strandflat surfaces, Lofoten Isles, northern Norway. Image credit: Sue Dawson.

rocks, characteristically 100–200m in width with backing cliffs up to 40m in height (Fig. 2.3). The cliff-platform junctions, together with the platforms themselves, do not show evidence of having been overridden by ice and thus are relatively recent features. The features are also distinctive in that many areas of platform occur in relatively sheltered fetch environments.

Shore platforms can also be produced by wave processes to produce wave-cut platforms. Many examples exist of shore platforms currently being produced along coastlines developed in less resistant rocks such as slates and mudstones. Because of the various problems associated with establishing the origins of shore platforms and backing cliffs, it has proved difficult to establish a relationship between the altitude of cliff-platform junctions and mean sea level. The majority of the scientific literature tends to ascribe the altitude of cliff-platform junctions to high water mark of spring tides. In general, the altitude of cliff-platform junction might provide an approximate indication of former sea level, but it can never be precise.

In numerous low latitude coastal environments, areas of planated rock have quite different origins. For example, many areas of coral reef are fringed by relatively flat surfaces that result from processes of biological construction (e.g. algal mat growth) rather than erosional processes. In these and other coastal areas, coastal features produced by biological construction may protect specific areas of coral and/or limestone from erosion, while elsewhere platforms may develop as a result of the sustained action of wave surf and salt spray. Along exposed coastlines with high wave fetch individual surfaces may occur up to several metres above mean sea level, while in more sheltered fetch environments and relatively tideless seas (e.g. the Mediterranean) the altitudes of individual surfaces may be no more than a few tens of centimetres above mean sea level.

Depositional shoreline features

Depositional indicators of past sea levels are located both above and below present sea level. Whereas submerged features are much more difficult to identify, uplifted features are common and range from marine and coastal sediments deposited under former high-energy conditions (e.g. storm beach ridges) to those deposited under quiescent conditions (e.g. mudflat sediments). For any uplifted depositional coastal feature to be of value in the reconstruction of past sea levels, it is necessary to assign a part of the landform or a **lithostratigraphic unit** in a coastal sediment sequence to a point on the former tidal cycle. For example, which part of an uplifted beach equates to the position of former mean sea level? In the following section, some of the most well-known coastal depositional landforms are described, and their altitude relationships with positions in the tidal cycle are explained. These relationships are often complex and difficult to interpret. For example, some uplifted coastal depositional landforms were produced in near-tideless seas (e.g. the Mediterranean) while others

Figure 2.3 Raised Late Pleistocene rock platform and cliff cut in limestone, Isle of Lismore, western Scotland. Image credit: Sue Dawson.

were deposited in areas of large tidal range (e.g parts of northern Europe and eastern Canada). Some features may exhibit surfaces obscured by hillslope material, fluvial sediments, and frequently by accumulations of blown sand. Others may have been subject to post-depositional changes that make them difficult to identify as clear features.

Storm beach ridges

Uplifted storm beach ridges are probably one of the most difficult landforms to relate to a former position of sea level. This is because the coarse debris of which a typical shingle ridge is composed may have been deposited many metres above the contemporary sea level (Fig. 2.4). Since high-energy storms can deposit shingle at different elevations, the altitude of a beach ridge tells us nothing about the former position of sea level except that contemporary sea level must have been at some point lower than the altitude of the ridge top.

Figure 2.4 Storm beach ridge, Pollochar, South Uist, Scottish Outer Hebrides, deposited inland during a major storm, 11 January 2005. Image credit: Sue Dawson.

Marine terraces

Marine terraces that occur above present sea level frequently occur in conjunction with a backing cliff (Fig. 2.5). Generally, such terraces decline in altitude seawards. They are distinct from shore platforms in that they are normally developed in unconsolidated sediment. It is common for such ancient shoreline features to be partially covered by hillslope sediments while alluvial fans may also occur on top of and adjacent to sections of terrace where backing cliffs have been dissected by streams. In many coastal areas, marine terraces are covered by a veneer of blown sand. Many studies of past changes in relative sea level have made use of the altitude of the cliff–terrace platform junction as an approximate indicator of former high water mark. However, along many

stretches of coastline where uplifted marine terraces exist, the precise location of the cliff–platform junction is obscured from view owing the presence of overlying sediment.

Uplifted and submerged coral reefs

In many low-latitude coastal environments where vertical land movements have taken place, staircases of Quaternary emerged coral reefs provide a chronology of past changes in relative sea level. In some areas (e.g. Sicily, Greece, Qatar, Philippines) the coral reefs extend hundreds of metres above present sea level (Fig. 2.6). One of the best-known studied sequences of emerged coral reefs is in the Huon Peninsula, Papua and New Guinea, where dating of individual coral species within reef complexes has made it possible to establish long-term patterns of relative

Figure 2.5 Raised marine terrace of Late Pleistocene age, Glenbatrick, western Jura, Scottish Inner Hebrides. Image credit: John Gordon.

Figure 2.6 Sequence of four elevated Pleistocene reef terraces, north coast of Bonaire, Dutch Caribbean. Image credit: Max Engel.

sea level changes. The flights of reefs extend up to *c.*600m above sea level and extend along the coast for over 80km. The emerged shore features are located in an area of active tectonic activity and in places are traversed by faults. In such areas dating of individual terraces in a 'staircase' (from highest to lowest at right angles to the coast) together with measurement of shoreline altitudes offers the potential for constructing detailed sea-level curves. However, this can only be accomplished if one is also able to measure past patterns of vertical tectonic uplift so that individual terrace altitudes can be adjusted to provide a more realistic chronology of past changes in relative sea level.

Submerged coral reef sequences occur in many low-latitude environments. In general, these descend to depths of between –100 and –120m and provide a record of relative sea level change since the last ice age. Two of the most studied sequences of coral reefs are from Tahiti and Barbados. Coral reefs fringe many volcanic islands. As Charles Darwin described, sinking of such islands enables lagoons to form between individual islands, leading to the development of barrier reefs. Continued island subsidence eventually results in the development of lagoons surrounded by coral reefs known as **atolls**. In these and other areas, sets of boreholes recovered

from sunken reef complexes have enabled the sampling and dating of coral species that are known to grow within specific water depth ranges. For example, the species *Acropora palmata* (Elkhorn coral) has frequently been used in sea level change reconstructions, since it generally lives at a water depth of *c.*6m with a vertical habitat range in the order of +/–2.5m (Fig. 2.7).

Other submerged evidence of former sea levels

In addition to drowned reef sequences, submerged shoreline features also occur in formerly glaciated areas, as well as in areas of tectonic subsidence. Field evidence for the occurrence of such

changes in relative sea level can range from buried peat deposits to areas of submerged forest (Fig. 2.8). Relative submergence can also take place as a result of land subsidence. Evidence for such submergence can range from the occurrence of drowned archaeological material (buildings, harbours, etc.) to sedimentary basins experiencing long-term subsidence. A submerged or buried tree stump or peat deposit does not by itself provide information on former sea level position apart from indicating that relative sea level must have been lower during the time of growth. Caution is needed in respect of coastal peat deposits, which in some cases may represent parts of former land surfaces that were later buried by blown sand. Coastal erosion and retreat thereafter results in the appearance, for example, of a coastal intertidal peat deposit that has nothing to do with former patterns of sea level change.

Figure 2.7 (above) Coral reef complex, American Samoa. Dating of individual submerged coral species can be used to reconstruct past patterns of relative sea level change. Image credit: NOAA/NFFS/PIFSC/CRED, Oceanography Team.

Figure 2.8 (left) Holocene tree stumps in present intertidal zone, Borth, south Wales. Image credit: Martin Bates.

Sedimentary evidence of former sea level changes

When scientists attempt to reconstruct past patterns of relative sea level change for a particular area, they often encounter situations where the morphological evidence for past changes is poor and the stratigraphic evidence is fragmentary. Frequently a single sediment exposure in the field is all that is available to reconstruct the former position of relative sea level. Under such circumstances a fundamental requirement is to demonstrate that the sediments in question are indeed of marine origin rather than having originated in different depositional environments. In some field settings, the presence within a sedimentary deposit of, for example, marine ostracods or sponge spicules will provide clear evidence of marine provenance of the sediment, but will not provide specific information on former sea level changes. However, the presence or absence of shallow-water marine or brackish fauna within a coastal sediment accumulation can provide more specific information on past changes in relative sea level (Fig. 2.9). Both marine macrofauna and microfauna are used extensively in such studies. Frequently the change from freshwater to brackish to marine sedimentation is represented by a change in the style of sedimentation. Sometimes the point at which there is a lithostratigraphic change (a change in the geological strata) is matched by a parallel change in the biostratigraphy (sedimentary strata distinguished by distinctive fossil assemblages). However, the situation often arises when, for example, a biostratigraphical change (e.g. from freshwater to marine environments) occurs out of phase with the change in the lithostratigraphy. This is an important issue since it influences the decision as to which stratigraphical units should be sampled for radiometric dating.

Figure 2.9 Sediment core sampling using the transfer function approach from a salt marsh, Nanortalik, western Greenland. Image credit: Anthony Long.

Marine microfossils
Foraminifera

Foraminifera are single-celled organisms with shells typically less than 1mm in size. Depending on species, the shells are most commonly composed of sand grains cemented together, organic compounds or crystalline calcium carbonate. Foraminiferal assemblages have been used in sea level change research to construct vertical zonation of saltmarshes according to tidal elevation and tidal range. These vertical zones have been used, in turn, to reconstruct past patterns of relative sea level changes using transfer functions where modern-day assemblages are used as analogues for former sea level conditions. Typically, specific sets of species of foraminifera occur within different elevation ranges within a specific saltmarsh. The different affinities of particular foraminiferal assemblages to parts of the tidal range then permit inferences to be made regarding fossil assemblages located in the same area, and in establishing their elevation relationship to past tidal regimes (Figs 2.9 and 2.10). This principle embodies the concept of the transfer function. It acknowledges that although elevation is the primary control on species distribution, other factors such as salinity, water temperature and grain size may also be important. In general, foraminifera have been used in sea level change studies in limestone areas and in other areas of carbonate bedrock where there is a sufficient supply of calcium for the construction of foraminiferal shell material. By comparison with morphological indicators of former relative sea level positions, the use of foraminiferal

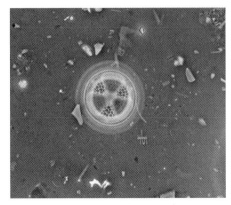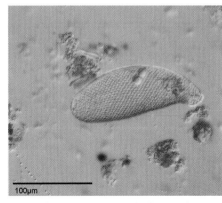

Figure 2.10 (Left) Cay foraminifera, Warraber Island, Torres Strait. Image credit: D. Hart. (Middle) Marine diatom *Actinoptychus senarius*. Image credit: Dave Ryves. (Right) Testate amoeba *Cyphoderia ampulla*, typically diagnostic of wet moss areas. Image credit: Picturepest, Creative Commons.

transfer functions to estimate former positions of relative sea level ranges is reasonably precise, with estimates of former tidal range positions having an accuracy of between +/−10–20cm.

Diatoms

Diatoms are algae capable of photosynthesis and are characterised by a siliceous skeleton. They typically range in size between 2 and 200 microns and are one of the most important groups of micro-organisms that occur in the world's oceans. In areas where the underlying geology provides an abundant supply of silica, diatom assemblages have been shown to exhibit a good relationship with tidal levels and tidal range. Many studies have shown that in salt marshes, particularly in coastal environments characterised by a high tidal range, diatom species assemblages correlate with different parts of the tidal cycle. Diatom assemblages can be grouped into distinct altitude ranges with diagnostic abundances of polyhalobous (those living in salt concentrations of > 30%), mesohalobous (0.2–30%) and oligohalobous (freshwater) taxa. In a similar manner to foraminifera, diatom-based transfer functions relate species assemblages to altitude and have been developed for specific coastal areas (Fig. 2.10). The transfer function is then used to reconstruct former tidal regimes for fossil assemblages, which, in turn, enables the construction of relative sea level curves for a given area.

Testate amoebae

Testate amoebae consist of a group of single-celled organisms known as protozoa. As the amoebae grow they produce a 'test' or shell typically in the order of 30–60 microns in diameter (Fig. 2.10). They occur in a wide range of environments including saltmarsh sediments. Their value in sea level change reconstructions is based on the belief that they are at least as precise as foraminifera and diatoms in reconstructing past changes in relative sea level. Testate amoeba assemblages are considered to exhibit distinctive intertidal vertical zonation across extensive geographical areas. As with sea level reconstructions using diatoms and foraminifera, transfer functions have been developed for intertidal testate amoeba assemblages. It is claimed that sea level reconstructions using this technique may have an accuracy as good as +/−10cm in some areas. Testate amoebae, together with foraminifera and diatoms that live in saltmarsh environments, provide valuable roles in helping extend records of former sea level change based on instrumental measurements further back in time.

Marine macrofossils

Marine macrofossils represent preserved marine organic remains large enough to be viewed with the naked eye. In many studies of relative sea level change, marine molluscs are routinely dated in order to help establish chronologies of past environmental change (Fig. 2.11). It should be noted, however, that in nearly all instances, the dating of marine mollusca provide almost

Figure 2.11 Marine fossil assemblage in uplifted Pleistocene marine terrace, Coelha, north San Nicolau, Cape Verde. Image credit: Ricardo Ramalho.

no information that can assist in the construction of curves of past relative sea level change. This is because the location and sampling points of individual marine mollusca have no association with a particular sea level – they provide no information other than demonstrating that a particular stratigraphic unit was deposited under marine conditions. Most marine mollusca have complex transportational histories prior to deposition, and there are very few mollusca that occur in growth positions.

3 Long-term chronologies of relative sea level change

Introduction

For over a century, while the debates on the causes of past sea level changes raged, field scientists continued to gather increasing amounts of evidence from around the world demonstrating past changes in sea level. The majority of this early sea level research took place across northern Europe, Scandinavia and North America. In Scotland as early as 1910, WB Wright was providing detailed descriptions of raised beaches in the Western Isles while similar detailed work was being published in Scandinavia by Gerard De Geer. Whereas there was agreement that the distribution of raised shoreline features in these areas was attributable to the effects of differential glacio-isostatic rebound, raised shoreline features located in lower latitudes were more difficult to interpret and explain.

Much of the scientific thinking on how global sea level may have changed worldwide was based largely on a four-fold scheme of Pleistocene glaciation. In order to understand this conceptual framework we have to go back over a century to the Alps, where in studies of glacial landforms and proglacial outwash terraces, Albrecht Penck and Eduard Bruckner argued that the Pleistocene ice age was characterised by four major ice ages, each separated by periods of interglacial warmth. They named the ice ages, from oldest to youngest, Gunz, Mindel, Riss and Würm. This idea of fourfold glaciation dominated the thinking of scientists for the first half of the twentieth century and came to be accepted globally (table 3.1).

Those who studied past changes in sea level also followed this path, assuming that each of the four glacial episodes was associated with periods of low sea level and that these were separated by interglacials during which global sea levels were high. Different names were assigned to the respective glacial and interglacial periods in different parts of the world (table 3.2).

Interglacials		Glacial stages
Monastirian	Upper: Variable Lower: Submergence to 60 ft	Würm glaciation Riss-Würm
Tyrrhenian	Upper: Variable Lower: Submergence to 60 ft	Riss glaciation Mindel-Riss
Milazzian	Upper: Variable Lower: Submergence to 190 ft	Mindel glaciation Günz-Mindel
Sicilian	Upper: Variable Lower: Submergence to 330 ft	Günz glaciation

Table 3.1 Classical scheme of fourfold glaciation based on Penck and Bruckner.

Oxygen Isotope Stage	USA		CANADA		USSR
1	Holocene		Holocene		European USSR
10,000					
2	Late Wisconsin		Late		Late Valdai
24,000		Wisconsin Stage			
	(35,000)	Late Pleistocene			
3			Middle		Middle Valdai
	Middle Wisconsin				
59,000					
4	*(65,000)*				
74,000	Early Wisconsin		Early		
5a	*(79,000)*				Early Valdai
85,000					
5b					
93,000					
5c	'Eowisconsin'		Sangamonian stage		
105,000					
5d					
117,000					
5e	*(122,000)*				
125,000	Sangamonian				Interglacial

Table 3.2 Nomenclature used for Late Pleistocene chronology of Canada, USA and Russia.

For example, in North America the last (Wisconsin) glaciation was preceded by the Sangamon interglacial and the earlier Illinoian glaciation. In the British Isles the last glaciation, known as the Devensian, followed the Ipswichian interglacial, which in turn, preceded the Wolstonian glacial stage and the earlier Hoxnian interglacial. By contrast, the last ice age across northern Europe was known as the Weichselian glaciation and was preceded by the Eemian interglacial.

The concept of four glacial periods of low sea level separated by interglacial high stands was well illustrated in Mediterranean sea level change research. Frederick Zeuner described for many areas flights of interglacial raised shorelines as (from oldest to youngest) – Sicilian, Milazzian, Tyrrhenian, Main Monastirian and Late Monastirian (table 3.1). Although we now know that most of the raised shoreline features across the Mediterranean are the result of vertical displacements caused by tectonic processes, the conceptual framework of four main periods of glaciation and low sea levels separated by intervening interglacial sea level highstands remained in place for many decades. This belief derived partly from interpretations of archaeological data, but was also based on stratigraphic relationships. For example, Zeuner maintained that in northern France, two Monastirian raised beaches are covered by wind-blown loess sediments that were deposited during the last period of general glaciation – thus these two raised beaches had to be older than the last ice age in Europe.

This particular example is cited here to illustrate how a sea level change

chronology was constructed prior to the advent of numerical dating techniques. Although it is now known that most of the raised shoreline features across the Mediterranean have been displaced vertically as a result of tectonic processes, the conceptual framework of four main periods of glaciation and low sea levels separated by intervening interglacial sea level highstands remained in place for many decades.

A challenge to the model of fourfold Pleistocene glaciation followed the publication of the Milutin Milankovitch astronomical theory of ice ages. Although the research of Milankovitch is principally concerned with past changes in global climate, it had major implications for the way that we understand past changes in sea level. Milankovitch presented a mathematical theory that explained how orbital

variations of Earth–Sun geometry led to changes in the past climate of the Earth. He developed a very detailed mathematical model that calculated latitudinal differences in solar insolation and inferred air temperature changes for the last c.600,000 years (Fig. 3.1). The work was paradigm shifting and appeared to demonstrate that the fourfold model of ice ages and intervening interglacials was too simplistic. The Milankovitch curve of past insolation changes appeared to indicate that past changes in global climate were of variable magnitude, irregular, and exhibited variations according to latitude. During the 1940s and 50s the Milankovitch theory received relatively little attention and was often criticised. The big change in the increasingly multidisciplinary study of sea level change came when scientists began to undertake

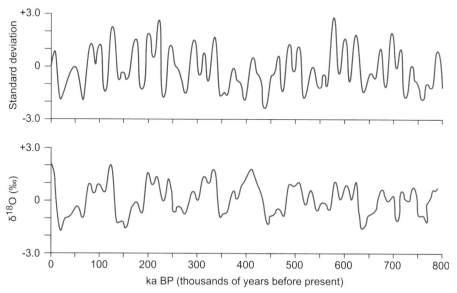

Figure 3.1 Generalised oxygen isotope curve for the world's oceans (bottom) compared with Milankovitch insolation curve (top). Image credit: adapted from J.J. Lowe and M.J.C. Walker (2014).

deep-sea drilling of ocean floor sediments and compared the results with the insolation curves of Milankovitch.

The marine oxygen isotope chronology

During the 1970s and 1980s, Nick Shackleton published several paradigm shifting papers on marine oxygen isotope stratigraphy of ocean floor sediments. His publications built on the pioneering research of Cesar Emiliani and laid the foundations of a new way in which to interpret past changes in global sea level. He argued that the composition of stable oxygen isotopes within the shells of foraminifera that have died and have been deposited on the ocean floor is partly due to the isotopic composition of ocean water when the foraminifera were alive, and it is also partly a function of the temperature of the seawater in which they lived. When evaporation takes place across the ocean surface, different isotopes of oxygen (^{16}O, ^{17}O and ^{18}O) in water molecules are released through the process of isotopic fractionation. The main changes in the isotopic composition of seawater mostly relate to changes in ^{18}O but they also partly reflect the water depth in which the shell carbonate is precipitated as well as the ocean water temperature in which they lived. With every lowering of ocean temperature there is a relative enrichment in the shell debris of ^{18}O compared to ^{16}O (Fig. 3.2).

Shackleton demonstrated how the measured oxygen isotope values in shells of benthonic foraminifera in ocean sediment cores varied with respect to a generally accepted standard value. In theory, during an ice age, ocean water is enriched in ^{18}O.

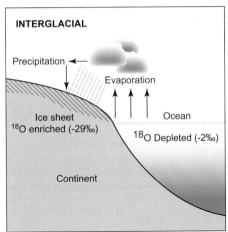
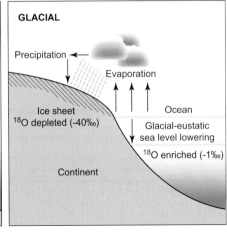

Figure 3.2 Respective enrichments and depletions of oxygen isotopes 16 and 18 across glacial and interglacial cycles. Image credit: A.G. Dawson (1992).

Conversely, atmospheric moisture that existed during an ice age and which was eventually deposited as snow on the world's major ice sheets is depleted in ^{18}O. During warm interglacial periods this process is reversed, with large amounts of ^{18}O removed from the world's oceans and a corresponding enrichment of ^{18}O in snow deposited on ice sheets (Fig. 3.2).

The role of ocean temperature changes

When Shackleton and co-researchers first measured the oxygen isotope changes in ocean floor sediment their analysis was based on the use of certain species of planktonic foraminifera. It was always understood, however, that planktonic foraminifera might only tell part of the story of past changes in global climate, since some ocean areas had experienced significant water temperature variations between interglacial and glacial periods. Attention gradually turned to parallel studies

of benthonic species of foraminifera, since ocean floor temperatures are just above freezing point and considered to have only changed slightly between glacial and interglacial periods. This implied that measurements of temporal changes in the ^{18}O content of benthonic foraminifera recovered from ocean floor sediments provide a much more accurate chronology of past changes in the isotopic composition of seawater. In other words, benthonic foraminifera provided a surrogate chronology over time of changes in the global volume of ocean water exchanged between oceans and ice sheets (Fig. 3.3).

If the greatest measured depletion of ^{18}O in benthonic foraminifera (the lowest point on the curve) from ocean sediment cores was known to be entirely due to a change in the isotopic composition of seawater, then it would be a simple matter to calculate the global ocean volume reduction during the maximum of an ice age. It would also be straightforward to convert such an

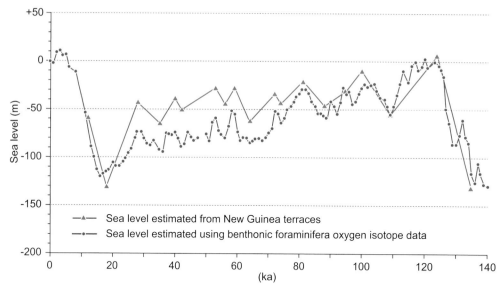

Figure 3.3 Oxygen isotope curve of inferred global sea level history for last *c.*140,000 years compared with pattern of relative sea level change reconstructed by John Chappell for New Guinea. The latter offset is due to the effects of long-term vertical tectonic movements. Image credit: A.G. Dawson (1992).

ocean volume loss into the volume of ice stored on land during, for example, the last glacial maximum. This is not entirely the case, however. The reason is that it has not proved possible to be sure if there is a component of oxygen isotope change within the shells of the sampled benthonic foraminifera that is due to changes in ocean floor water temperature – albeit a small change.

Scientists still debate vigorously about possible mismatches between the volume of global ice during the last glacial maximum and the volume of ice estimated from oxygen isotope studies of ocean floor sediments. Nevertheless, Shackleton's pioneering curve of global ocean volume change over the last *c.*150,000 years based on ^{18}O benthic foraminifera data is regarded as the best graph of glacio-eustatic

changes in sea level that we have (Fig. 3.3). Since then, much more detailed analyses have been undertaken, also covering longer periods of geological time – all studies have served to confirm Shackleton's hypothesis. Not only that, but the oxygen isotope chronology derived from ocean floor sediment is, in its broad outline, similar to the Milankovitch curve of solar insolation changes (Fig. 3.1). This appeared to indicate that the cycles of global ice volumes changes, and hence glacial–interglacial cycles reconstructed from the marine oxygen isotope record may have their origins linked to past changes in solar insolation linked to long-term orbital changes of the Earth around the Sun. Recognition of the validity of the marine oxygen isotope record as an indicator of past changes in the

Earth's climate (including sea level changes) ultimately led to the abandonment of the model of four major episodes of Pleistocene glaciation. In terms of sea level change, the new findings were pivotal, since it became clear that long-term changes in the volume of water stored in the ocean could be demonstrated to behave in an approximately synchronous manner with long-term glacial–interglacial cycles.

4 Reconstructing past changes in relative sea level

Introduction

In order to understand modern-day patterns of relative sea level changes for specific areas and to try and predict future changes, it is essential to have as complete a knowledge as possible of past patterns and sequences of change that have taken place over different periods of time. Over the last century during which the majority of field investigations have been undertaken, scientists have faced two particular problems. The first has been the need to associate individual pieces of field evidence for past relative sea level positions to particular sea level elevations. The second has been the need to establish accurate ages of former sea level positions. In this section a brief summary is presented for the principal methods that are used to reconstruct past patterns of relative sea level for given areas. This is followed by a short summary of the key dating methods which are used, together with a brief consideration of the limitations that one should be aware of when try to establish past patterns of relative sea level change for particular areas of the Earth's surface.

Sea level curves

The production of a former sea level curve for a particular area is difficult to achieve. The plotting of individual curves is based on an internationally agreed methodology whereby former sea level positions are reconstructed making use of data from low-energy sedimentary environments. Individual points plotted on any curve are known as sea level index points. The requirements for individual sea level points to be considered valid include the need for a sediment sample to have a known location, an age as well as an established relationship between the sample and a known tidal position. The sample also needs to have an indicative meaning that is defined by a reference water level (for example, does the sample indicate former high water mark or former mean tide level, etc.) and also that it has a stated indicative range that simply refers to the altitude range within which the sample could possibly occur.

The construction of sea level index points also includes analysis to determine whether individual samples exhibit a sea level tendency indicating either an overall increase in marine influence (a relative marine transgression) or an increase in freshwater influence (a relative marine regression). These techniques, developed by Michael Tooley and co-researchers and latterly made popular by Ian Shennan at the University of Durham, are now used worldwide. Their application is best illustrated by a very simple example. In a particular coastal area a sediment exposure shows peat overlain by marine silts and clays containing marine mollusca. The lithostratigraphy thus points to change from freshwater to marine sedimentation at the stratigraphic transition between peat and the overlying silt. However, the diatom biostratigraphy does not point to the same boundary position as indicative of the transition from freshwater to marine sedimentation. The sea level index point for this location is thus selected as the point where the transition between oligohalobous diatom species and polyhalobous species occurs, and not where there is a change in the lithostratigraphy. A sediment sample from the point in the sediment core where the diatom change takes place is then identified as a sea level index point and submitted for radiometric dating. In this example, the sea level tendency is that of a relative marine transgression. By contrast, if one encountered lake sediments resting upon marine and brackish sediments, the likelihood would be that the sea level tendency would be that of a relative marine regression.

Isolation basin studies

Another well-known method of sea level reconstruction is through the use of coastal isolation basins (Fig. 4.1). These

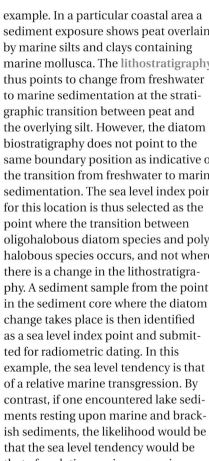

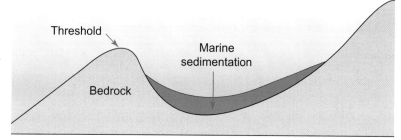

Sea level 'A'

Threshold

Marine sedimentation

Bedrock

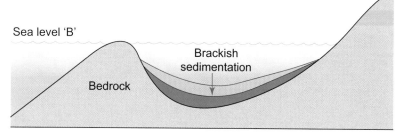

Sea level 'B'

Brackish sedimentation

Bedrock

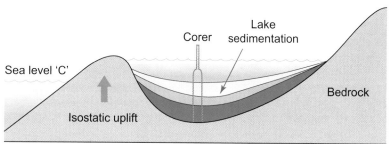

Sea level 'C'

Corer

Lake sedimentation

Bedrock

Isostatic uplift

Figure 4.1 Illustration of the Ulf Hafsten model of use of isolation basins in order to reconstruct past patterns of relative sea level change. Image credit: adapted from D.E. Smith and A.G. Dawson (1983).

are topographic depressions that have in the past been submerged by the sea, or have been isolated from it by past changes in relative sea level. The depressions are for the most part controlled by local geology but in some cases they are formed in glacial sediments.

The use of isolation basins in sea level change research, first popularised by Ulf Hafsten in Norway, has mostly been undertaken in areas affected by glacio-isostatic rebound. Each isolation basin should preferably have a rock lip at its seaward margin. Some isolation basins are presently freshwater lakes, in which case lake sediment coring is undertaken to determine whether or not the lake has been inundated by seawater at any time in the past. If marine inundation of the basin has taken place, stratigraphic analysis of the lake sediments could be undertaken in order to determine the age when this took place. Similar methods could also be used to determine the timing of periods of relative marine regression that isolated the basin from the sea. In examples such as these, tendencies of relative sea level change can only be established with reference to the altitude of the rock lip at the seaward margin of the basin. For example, imagine an isolation basin lake with a rock lip at 20m above mean sea level. Sediment coring in c.30m water depth recovers a core that, after laboratory analysis, shows limnic sediments resting beneath marine sediment with a relative marine transgression dated to 6500yr BP. In an example such as this, the location of the lowest point of the rock lip becomes the sea level index point location and demonstrates the occurrence of a former relative marine transgression that took place at this time at an altitude of +20m.

This example is used here to demonstrate several important issues. First, the determination of the rock lip altitude is critically important. The lowest point across the rock lip needs to be known, since it is this altitude that determines when the lake was inundated by the sea, and latterly when the sea receded from the lake. In addition, one needs to be sure that the rock lip is in fact a bedrock feature. If sediment overlies a rock lip and obscures bedrock, the measurement of the lowest point may prove to be false owing to the possibility that the sediment lip may have been subject to later downcutting by streams.

If several isolation basins occur in a 'staircase' of elevation, their investigation provides the opportunity to develop a relative sea level curve for that location, providing the entire 'staircase' sequence of isolation basins are cored and analysed. In any such study, the altitude differences of the various rock lips should be as narrow as possible. This is because the only sea level index points that can be determined are in respect of each rock lip location and altitude. In this way in a 'staircase' of, say, seven isolation basins may provide evidence of the same relative marine transgression having taken place in the first five basins but not in the highest two. The rate and timing of the relative marine transgression provides five sea level index points with five different ages that correspond to the altitude of the five rock lips, while the upper limit of the relative marine transgression lies somewhere between the altitudes of rock lips five and six.

The stratigraphy of individual isolation basin lakes is therefore complex. When taking sediment cores from isolation basin lakes, great care must be taken to sample from the deepest part of the basin and avoiding locations where sub-aquatic slumping may have taken place. Ideally, such studies need to be preceded by a bathymetric survey to determine the most suitable location for lake sediment coring. Some isolation basins can be sampled by hand-operated corers on dry land. Such basins are those that have infilled with vegetation over time and have become a hydrosere.

Stratigraphy of estuaries and deltas

A common method of sea level change reconstruction is to undertake stratigraphic research in estuaries and deltas. In these areas, significant deltaic sedimentation is coupled with subsidence and has continued over thousands of years. The local stratigraphy is therefore highly complex with fluvial sedimentary units inter-fingered with coastal and marine sediments. All deltas are subject to rapid lateral migration of distributary channels. Such changes, when considered in conjunction with changes in relative sea level, river flooding and marine inundation during storms, result in highly complex patterns of sedimentation. For these reasons, in the largest delta complexes (e.g. Mississippi and Nile), it is often difficult to correlate boreholes across large areas. In estuaries, however, where boreholes can be more closely spaced, correlation of stratigraphic units between cores is much easier. It is not surprising that some of the classic studies of Holocene sea level reconstruction have been undertaken in such areas, for example, the well-known research of Michael Tooley for Morecambe Bay, NW England; the pioneering sea level change research of Saskia Jelgersma in the Netherlands, and the classic work of Brian Sissons and his team from the Forth Valley, Scotland (Fig. 4.2). This latter work illustrates

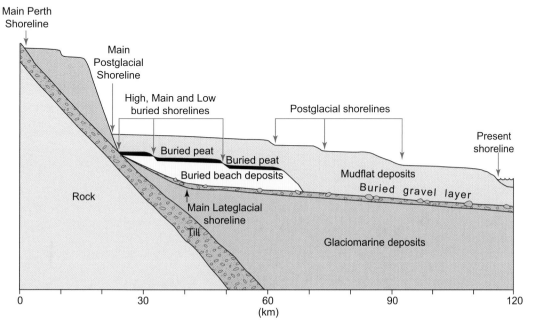

Figure 4.2 Brian Sissons' classic representation of surface and buried morphology as well as stratigraphy of upper Forth valley, eastern Scotland. Image credit: adapted from J.B. Sissons (1967).

perhaps best of all the methodology that underpins this type of research.

Sissons initially mapped at a scale of 1:10,560 (6 inches to a mile) all surface landforms of marine, estuarine, fluvial and glacial origin. Over many years he then undertook instrumental levelling to centimetre accuracy of all shoreline features using a grid spacing of *c.*50m. This provided a network of *c.*12,000 altitudes of both buried and emerged shoreline features. He then completed over 2500 boreholes using hand-operated coring equipment, this supplemented by a similar number of commercial borehole records. Thereafter numerous pollen and diatom studies were undertaken to identify sea level index points and obtain age estimates through the use of radiocarbon dating. Studies such as this (albeit in most instances not at this level of detail) have been undertaken across the world in order to construct curves of local relative sea level change. They have the advantage over isolation basin studies, of enabling buried shorelines to be traced over many kilometres (Fig. 4.2).

Shoreline reconstruction in areas of glacio-isostatic rebound

In formerly glaciated environments such as eastern Scotland that are characterised by glacio-isostatic rebound, coastal landforms and sediments are gradually uplifted above contemporary sea level provided that the rate of crustal uplift exceeds the rate of any rise in ocean level. Coastal features formed in this way are uplifted differentially on the basic premise that the absolute amount of uplift will be greatest in areas where the ice was formerly thickest and at progressively lower elevations across areas where the ice was formerly thinner. When raised shoreline features of approximately the same age are plotted on a map, the reconstructed pattern reveals a broadly concentric ring-shaped pattern of shoreline deformation around the centre of a greatest vertical rebound (i.e. where the ice was formerly thickest) (Fig. 4.3). However, for areas where the last ice sheet was characterised by multiple ice domes and saddles (e.g. the Laurentide ice sheet) the geometry of uplift isobases for a given shoreline of known age may exhibit much more complex patterns of uplift.

In any formerly glaciated environments, the distribution of raised shoreline features is usually very fragmentary. Having recognised and mapped a raised marine terrace, one might not find another similar feature presumed to be of the same approximate age for several kilometres. Elsewhere, one might find 'staircases' of raised beach ridges in a particular area and nothing in nearby coastal areas. The latter circumstance occurs, for example, in parts of northern Canada and Alaska, where isolated areas of unvegetated beach ridges occur in staircases (Fig. 4.4). If several of the beach ridges contain marine shells suitable for dating, it is possible to produce an approximate relative sea level curve

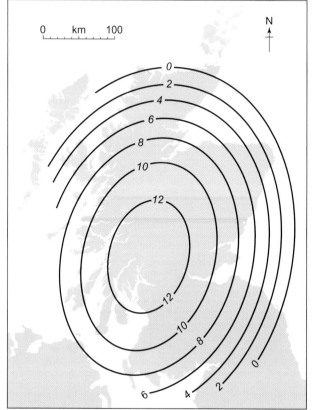

Figure 4.3 Example from Scotland of concentric shoreline uplift isobases around a single dome of glacio-isostatic rebound. Image credit: S. Dawson and D.E. Smith.

Figure 4.4 Holocene raised beach ridges, Cape Krusenstern, Kivalina, NW Alaska. Image credit: Creative Commons, Education Specialist.

for a single location. However, since these circumstances are very rare, it is often necessary to undertake painstaking mapping of raised marine features across hundreds of square kilometres and the radiometric dating of as many of these features as possible.

The key problem that remains for such landscapes is how to correlate one raised shoreline fragment with another, given that they may sometimes be many kilometres apart. Since individual raised shorelines of the same age exhibit a decline in altitude with increased distances away from the centre of isostatic uplift, one cannot correlate raised shoreline features on the basis of similar altitudes. By dating as many individual raised shoreline features as possible, raised shoreline features at different altitudes can be grouped together and correlated with each other on the basis of age.

Provided that individual raised shoreline features can be correlated and shown to be of the same approximate age, shoreline height–distance diagrams can be produced (Fig. 4.5). Such diagrams are schematic representations of individual raised shorelines of given ages plotted at right angles to the shoreline uplift isobases. In general, the oldest raised shoreline features exhibit the greatest amounts of glacio-isostatic tilting, with younger shorelines plotted at right angles to the uplift isobases exhibiting decreasing amounts of tilting. Thus a raised shoreline in western Norway dated to *c.*16,000 years ago

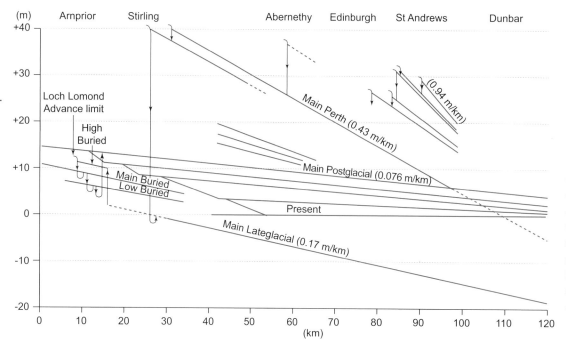

Figure 4.5 Example of shoreline height–distance diagram for the Forth valley, eastern Scotland. This diagram, produced by Brian Sissons over 50 years ago, shows how raised shorelines in an area affected by glacio-isostatic uplift decline in altitude away from the centre of uplift. The oldest shorelines exhibit the greatest amounts of isostatic tilting. Image credit: adapted from J.B. Sissons (1967).

may exhibit a regional tilt away from the centre of uplift in the order of, say, 1.5m/km whereas a raised shoreline of the same approximate age in Scotland may have a regional tilt at right angles to the uplift isobases of 0.4m/km.

Apparent age of seawater and isotopic fractionation

Radiocarbon dating of a modern marine mollusc on any beach will give an 'old' age as a result of the 'apparent age of seawater'. For example, along coastlines bordering the North Atlantic the ages of modern marine shells typically produce radiocarbon ages of *c*.400–440 years BP (before present). In other marine environments the value for apparent age is quite different. For example, along the Mediterranean coastline, the apparent age of seawater is in the order of *c*.60–80 years while in Antarctic waters apparent ages often exceed 1000 years. This counter-intuitive observation that seawater has an apparent age is due to the circulation of carbon in the water. The calculation of the apparent age of seawater for different regions of the world has been established by sampling numerous modern marine shells and dating them by the radiocarbon method. For some coastal environments, e.g. NW Europe, the apparent age is calculated by determining the average age of hundreds of samples. In other oceans, the estimated apparent age is based on the dating of a small number of shells. The issue is an important one, since the age determination of a marine mollusc in a raised beach deposit has to have the apparent age subtracted from the measured age. Matters become even more complicated since for certain ocean areas, the apparent age of seawater has varied over time. For example, in northern Europe the apparent age of seawater during the Lateglacial period was nearly double its modern value. An additional complicating process is isotopic fractionation where marine carbonate sediments exhibit greater radiocarbon activity than terrestrial organic material. This process can cause marine sediments to appear preferentially 'younger' than they actually are.

Comparing radiocarbon and sidereal timescales

Radiocarbon dating of a sea level index point has to contend with the fact that the radiocarbon timescale does not correspond precisely with real time age estimates. Historically the radiocarbon timescale has been calibrated by linking it with the chronology obtained from the study of tree rings (dendrochronology). Since dendrochronology covers the last *c*.9000 calendar years, this mean that radiocarbon dates can only be calibrated to a sidereal year timescale as far back as they can be calibrated against a tree ring chronology. This problem has partly been resolved in recent years through the parallel dating of fossil coral reefs species by mass spectrometry dating using uranium–thorium ratios on the one hand and by ^{14}C on the other. Edouard Bard and his team analysed a set of samples of the coral *Acropora palmata* older than *c*.9000yr BP, used by Richard Fairbanks to construct a relative sea level curve for Barbados. They found that the ^{14}C ages were consistently older than the $^{234}U – ^{230}Th$ for the same samples, reaching a maximum separation in age of nearly 3500yr close to the last glacial maximum. Two possible reasons exist to explain this discrepancy. First, it could be due to temporal exchanges in the ^{14}C reservoir between atmosphere and ocean. The preferred explanation by Bard, however, is that the Earth's magnetic field may have changed over this time interval, causing changes in the nature of the shield that enables the production of ^{14}C by cosmic rays. This issue is particularly important since dating of the Barbados submerged coral reefs species by $^{234}U –^{230}Th$ results in a quite different relative sea level curve from that resulting from ^{14}C dating (for a thorough review of radiocarbon dating issues and timescale calibration see Lowe and Walker, 2014).

These descriptions thus serve as a salutary reminder that extreme caution needs to be exercised when attempting to interpret past patterns of relative sea level change for individual field locations. One must not forget, also, that the aforementioned dating techniques each have their own limitations, some of which would have been unknown to scientists working several decades earlier. The submerged coral reef research illustrates this issue very clearly, since it could not have taken place without the development of the technique of uranium–thorium dating. It will doubtless be the case that in the future, new sea level change dating techniques will be developed and refinements in the construction of individual relative sea level curves will be achieved.

5 Response of Earth's crust to surface loads

Introduction

It might seem surprising to include a discussion on geology within a textbook on sea level change. Nevertheless, some key geological processes lie at the very heart of understanding how we interpret some of the key issues surrounding the debate on sea level change. Why should this be? The reason is that during the glacial–interglacial cycles of the Pleistocene the Earth's crust has responded in various ways to the loading and unloading of ice sheets, as well as to changes in ocean volume. But none of these changes can happen within the lithosphere and mantle without some form of compensatory movement of subcrustal material. Indeed, the movement of tectonic plates, susceptible themselves to sudden movements, depend upon transfer of material at depth beneath the Earth's crust.

Consideration of these issues raises many questions. What happens to the Earth's crust, for example, over timescales of thousands of years when the melting of huge volumes of ice on land increases the volume of water stored in the world's oceans? How do these changes manifest themselves? The most obvious starting place in sea level change research is to understand how and why the Earth's interior possesses an elasticity that allows it to respond to such changes.

The Earth's crust and mantle refer to layers of the Earth's interior that are defined in terms of their chemical properties. By contrast, the lithosphere and asthenosphere represent internal zones of the Earth that are defined in terms of their mechanical properties. The lithosphere effectively represents the tectonic plates that move over a much weaker and deformable asthenosphere. The asthenosphere is thus crucially important, since it enables tectonic plates to move. The Earth's crust is distinctive in terms of its chemical properties and contains many of the lighter elements, e.g. silicon, oxygen, calcium, potassium, sodium, aluminium, etc.

Ocean crust is typically thin, c.5–10km, whereas continental crust is thicker, typically 30–35km. The underlying mantle is similar to the crust in the sense that it is composed of significant quantities of silicon and oxygen, but it also contains much more iron and magnesium. The mantle extends downward to c.2900km towards the boundary with the Earth's core. Within the mantle occurs the lower part of the lithosphere, the asthenosphere as well as the mesosphere (Fig. 5.1).

In 1914, Joseph Barrell first introduced the words 'lithosphere' and

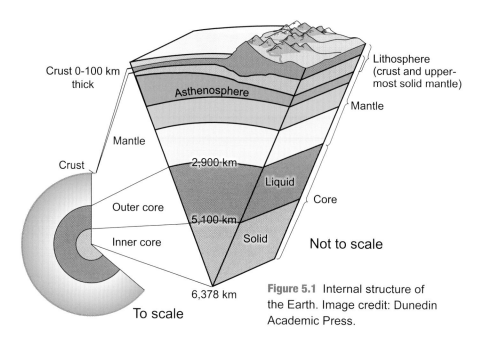

Figure 5.1 Internal structure of the Earth. Image credit: Dunedin Academic Press.

'asthenosphere' to describe the 'strong' outer crust of the Earth overlying a weaker layer. He put forward the view that deeper-seated material must be able to move beneath the Earth's crust in order to compensate for the lateral movements that take place across the Earth's surface. The asthenosphere is thus considered today as a mechanically weak region of the upper mantle occupying a region typically between *c.*200–700km depth. Many regard the boundary between the base of the lithosphere and the top of the asthenosphere as coincident with the 1300˚C isotherm. Above this level the lithosphere is effectively rigid, while below it is viscous. It is this viscosity that enables deformation to take place. Rates of flow are typically in the order of cmyr⁻¹ within huge convection currents in the order of thousands of kilometres. For this reason scientists frequently refer to the rigid lithosphere as 'floating' over a slowly moving asthenosphere – thus enabling the movement of tectonic plates.

The precise viscosity of the asthenosphere is the subject of considerable debate – is it uniform everywhere or does it vary regionally? Furthermore, if the viscosity of the asthenosphere is too low plates will move too fast; if it is too high they will not move at all. So what is viscosity? It is a measure of the internal friction of a fluid that makes it resist a tendency to flow. A material that needs a shear stress of one dyne (a unit of force equal to 10^{-5} Newtons) per cm^2 to produce a shear rate of 1 second has a viscosity of 1 poise. In geology, viscosity is often described in terms of Pascal-seconds (Pa s) where each Pascal-second is

equal to 10 poise. If material in the asthenosphere behaves as a Newtonian fluid, it is assumed that there is a linear relationship between shear stress and shear rate. Under such circumstances the viscosity of the fluid remains constant as the shear rate changes.

Generally speaking, material in the asthenosphere deforms through a combination of diffusion and creep. Some regions of the asthenosphere are characterised by much lower viscosities than others. In particular, there is general agreement that there is a relatively thin low-viscosity region at the base of the upper mantle. The asthenosphere is therefore separated by strong rock both beneath and above it. Above is the upper part of the mantle and the crust, while below it is the relatively strong solid mantle. Seismic velocities tend to exhibit a progressive increase the deeper one descends into the mantle as a result of increased rock pressure and density. At depths generally between 100km and 250km seismic wave properties appear to indicate the presence of a partly molten zone considered indicative of the asthenosphere. Another seismic discontinuity occurs at a depth of 660–670km. Here there is a sharp jump in seismic velocities, generally interpreted as marking the boundary between the less dense upper mantle and the denser lower mantle.

Viscosity of the asthenosphere and sea level change

Apart from the key issues surrounding the nature of climate change, the measurement of patterns of sea level change worldwide is of great importance to the discipline of geology, since the data can be used to provide valuable information

on the nature of asthenosphere viscosity and our understanding of plate tectonic movement. Equally, estimates of asthenosphere viscosity can be used to learn more about the ways in which the Earth's crust has responded to the melting of ice sheets worldwide following the last glacial maximum and the transfer of huge volumes of water into the world's oceans. Since these adjustments are still taking place today, the way in which the asthenosphere behaves plays a fundamental role in our understanding of the nature of sea level changes in the decades and centuries ahead. But how do we calculate the viscosity of the asthenosphere?

Dick Peltier has described this issue as attempting to solve 'the inverse problem for mantle viscosity'. By this he refers to the compilation of a subset of reliable and accurately dated indicators of past relative sea level changes that can be related to parameters of asthenosphere viscosity capable of producing a numerical model that can replicate patterns of past sea level change. Peltier and co-researchers found that data on past relative sea level changes could be 'tuned' to a numerical model characterised by a well-defined increase in viscosity at the 660–670km seismic discontinuity where the values jump from *c.*0.4×10^{21}Pa s to *c.*1.2×10^{21}Pa s. Of course, different models give different viscosity parameters. Separate modelling undertaken by Kurt Lambeck and his team considered that the 'jump' in viscosity across the 660–670km discontinuity is much higher than that described by Peltier. This is not the place to discuss the finer details of the pros and cons of the various models. Suffice it to point out here that asthenosphere viscosity is one of the most important factors in reconstructing past patterns of relative sea level change. We need only remind ourselves that a very low viscosity asthenosphere will always be associated with lithospheric plates that are mobile and extremely sensitive to loading and unloading by ice and water. By contrast, if the asthenosphere viscosity is pushed to too high a value, the lithospheric plates will struggle to move. As we shall see, this 'chicken and egg' problem of tuning the profile of mantle viscosity to the relative sea level change data keeps recurring in various forms when we try to interpret present-day rates of relative sea level changes and predict future changes.

Introduction

In the debate on sea level change, an enduring theme is the fundamental difference between processes of relative sea level change caused by changes in ocean mass, and those caused by changes in the volume of the world's oceans. In general, changes in ocean mass are related to increases in ocean volume caused by melting ice. By contrast, the most important process of ocean volume change is that due to the heating of the world's oceans. Both processes are linked to climate change, with many arguing that since ocean mass increases are linked to melting ice there has to be a clear connection to global warming. A similar viewpoint can be taken in respect of ocean volume change, since blame for the additional heat stored in the world's oceans could also be placed at the door of recent climate change. As noted earlier, relative sea level changes are also responsive to deformation of the lithosphere arising from changes in the distribution of loads of water and ice across the Earth's surface. These changes are underpinned by two key concepts in sea level change research – 'eustasy' and 'isostasy'.

In the scientific literature the term 'eustasy' is often used to describe a state of equilibrium in the shape and level of the surface of the world's oceans. The term derives from the Greek 'eu' meaning 'well' or, in an earlier interpretation, 'to be' and 'statis' for the act of standing or being in balance. Separately, the term 'isostasy' refers to the state of equilibrium that exists between the Earth's lithosphere and the underlying mantle, such that the crust essentially 'floats' on the underlying material. The term 'isostasy' derives from the Greek, 'isos' meaning equal and 'stasis' (as in 'eustasy') referring to balance. Changes in ocean volume where there is no change in global ocean mass are referred to as steric changes. Although such changes include changes in the sea surface that take place during storms and hurricanes and as a result of changes in ocean circulation, the most important steric change relevant to the sea level change debate is that associated with the thermal expansion of ocean water. Each of these three processes (eustasy, isostasy and steric change) is described below. As will become evident, some eustatic processes incorporate elements of both ocean mass and ocean volume change. By contrast, isostatic processes relate to the lithospheric response to changing ice and water loads. Separately, steric processes are associated with ocean volume changes.

◆ Glacial eustasy: the process by which water is exchanged between the world's glaciers and ice sheets and its oceans, leading to worldwide changes in sea level.

◆ Geoidal eustasy: the process by which seawater is distributed across the world's oceans as a result of gravitational processes.

◆ Tectono-eustasy: processes by which ocean water is displaced relative to the land by crustal movements.

◆ Glacial isostasy: processes by which the lithosphere is depressed due to the weight of an overlying ice sheet. It also includes the crustal rebound associated with the removal of an ice cover.

◆ Hydro-isostasy: isostatic deformation of the ocean floor as a result of changing water loads.

◆ Steric changes: ocean processes that involve changes in ocean volume but not ocean mass (including effects caused by changes in ocean temperature, density and salinity, as well as changes in ocean circulation).

Eustasy

In sea level research, eustasy refers to the global sea surface being in balance, or equilibrium, with the Earth's geoid. Eustasy refers to global rather than regional or local change, and thus a eustatic change could refer to a global change in the amount of water stored in the world's oceans (an ocean mass change). Equally it could refer to an absolute change in the volume of the ocean basins. Implicit in this definition of eustasy is that ocean surfaces occur in gravitational balance with the gravitational field of the Earth. Changes in the rise and fall of these surfaces can be the result of transfers of water mass into ice due to the melting or build-up of glaciers and ice sheets on land (referred to as glacio-eustatic changes). Eustatic changes can also take place due to spatial changes in gravitational attraction of ocean water that result in changes in the shape of the equipotential ocean surface (known as geoidal-eustatic changes). This presents a difficulty for the scientist, for example, who has measured and dated a raised shoreline feature in the field. The problem is how to establish how much of the measured change in relative sea level in the field is due to past glacio-eustatic changes, and how much of the measured change is due to geoidal gravitational changes. The only way that these two processes can be quantified and distinguished from each other for a specific field location is through the application of geophysical modelling where patterns of former relative sea level changes can be modelled for the whole Earth. Fortunately, a group of distinguished geophysicists led by Dick Peltier, Kurt Lambeck, Jerry Mitrovica and Glenn Milne have made huge progress in this regard.

Eustatic changes can also take place as a result of changes in the shape of ocean basins, these being known as tectono-eustatic changes. It should be noted that these different types of eustatic process interact with each other. For example, ocean water is gravitationally attracted to

ice sheets – meaning that any changes in the mass balance of an ice sheet can lead to changes in the geoidal sea surface. Thus a eustatic change in sea level can be caused by the combined effects of a global change in the mass of water stored in the world's oceans together with gravitational changes, thus making it difficult to separate one effect from another. In the scientific literature several categories of 'eustasy' are described:

Glacial eustasy

Glacio-eustatic changes take place as a result of the growth and decay of ice sheets, ice caps and glaciers. Long-term glacio-eustatic changes are normally described in terms of Pleistocene glacial–interglacial cycles. During warmer interglacial periods, less ice was locked up in glaciers and ice sheets, and as a result there was a global rise in glacio-eustatic sea level (greater ocean water mass). During

glacial periods, the growth and expansion of glaciers and ice sheets led to a glacio-eustatic lowering of sea level (reduced ocean water mass) (Fig. 6.1). At shorter timescales, global ocean mass can also change if water is added to (or abstracted from) the ocean from other areas where water is stored. In the Earth's climate system these include reservoirs, groundwater and through other ways in which humans alter the land surface and modify the hydrological cycle.

It is important to know how much glacio-eustatic sea level lowering took place during the last glacial maximum. If this value is known for different locations, then it becomes possible to estimate the volume of ice locked up in the last great ice sheets. If such a figure can be calculated it can then be compared with estimates of how much ice existed in different areas of the world during the last glacial maximum. In theory this figure should match the ice

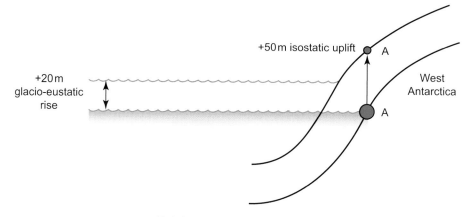

Relative sea level fall 'A' = -30m

Figure 6.1 Illustration of interaction of glacial eustasy and glacial isostasy. The example shows how +50m of glacio-isostatic rebound across West Antarctica, accompanied by a glacio-eustatic sea level rise of +20m, results in a relative sea level fall of −30m.

volume estimate reconstructed on the basis of estimates of glacio-eustatic lowering for the last glacial maximum. An additional difficulty is that during the peak of the last ice age, different ice sheets may have reached their maximum extents at different times.

The calculations of former ice sheet volumes are fraught with problems. The greatest unknown is in respect of the largest ice sheets – namely those that covered Antarctica, North America, Scandinavia and Russia. Many sections of these ice sheets extended far beyond the present-day coastline. Thus, the former extent of ice is often represented by ice-marginal moraines that are presently submerged. Not only is it necessary to identify and date such moraines in order to be sure that the landforms were in fact deposited during the last glacial maximum (rather than during any later or earlier time interval), but it is also important to be able to map and date such moraines across huge areas of seabed in order to be sure of the former spatial extent of each former ice sheet. In the case of Antarctica, much of this mapping and dating has not yet been done.

It is also necessary to know the thickness of each of these former ice sheets if one is to make accurate statements concerning volumes of ice locked up in these former ice sheets. For example, how are we to estimate the thickness of the Laurentide ice sheet in North America, given that the vast majority of the landscape was buried by ice, and no bedrock outcrops (nunataks) protruded above its surface? The only way to make such an estimate is to use glaciological models that replicate a cross-profile of an idealised ice sheet. Thus the estimates of the volumes of ice locked up in the Antarctic, Laurentide (North American), and Fennoscandian (Eurasian) ice sheets during the last glacial maximum are based on a combination of geo-morphological mapping, radiometric dating and mathematical reconstructions of former ice sheet dimensions.

Aware of these difficulties, several scientists including Richard Fairbanks and Eduard Bard have attempted to make estimates of glacio-eustatic sea-level lowering for coastal areas far removed from the last great ice sheets. Their principal areas of interest have been Barbados and Tahiti. For these islands the research focus has been submerged coral reefs that enabled the construction of records of past relative sea level change. Since particular species of coral, of which *Acropora Palmata* is the most common, grow in a specific depth of water below the sea surface (in this case at depths of 5–6m), it is possible to sample and date submerged coral reef complexes in order to determine former positions of sea level at specific times in the past. In both of these areas, the deepest coral reef complexes are located at water depths of *c.*–120m, and hence it has been concluded that at least for both of these areas, coral reefs have been produced at this approximate water depth since at least the maximum of the last ice age and possibly earlier.

Tectono-eustasy

Tectono-eustatic changes are those processes by which the volume of ocean basins is changed mostly as a result of changes in the volume of ocean ridge systems and as a result of marine sedimentation. Over geological timescales, plate tectonic processes result in changes in the shape of ocean basins. Although such tectono-eustatic changes take place at timescales of millions of years, they can also take place suddenly as a result of earthquake activity (Fig. 6.2). In the case of the long-term evolution of ocean basins, tectono-eustatic changes in the past have resulted in significant changes in the dimensions of oceans, and hence sea levels.

Throughout the Earth's history, the growth of ocean ridges and changes in patterns of ocean floor sedimentation has led to the development of wider oceans and a lowering of overall sea level. By contrast, during times when the oceans were generally narrower, sea level would have been higher. Since the drifting of geological plates takes place at rates of cmyr^{-1}, such tectono-eustatic changes have had a negligible effect on the pattern of global sea level change during Pleistocene glacial–interglacial cycles, and especially during the Late-Pleistocene and Holocene epochs. Other tectonic processes associated with vertical movements of the Earth's crust have also had a dramatic effect on sea level throughout the Earth's geological history. These have included mountain-building processes arising from the collision of oceanic and continental plates, as well as within island arc provinces. They have also involved crustal processes associated with spreading centres and the evolution of volcanoes over hotspots. These processes also range from slow epeirogenic movements to sudden earthquakes.

Given the above, one might

Figure 6.2 Emerged shoreline notches, Milokopi Bay, Perochora peninsula, eastern Gulf of Corinth, Greece. The uppermost strandline is of middle-Holocene (*c.*6kyr BP) age. Multiple minor notches occur below this level and are due to repeated seismic uplift events, the most recent of which was the 1981 Corinth earthquake. Image credit: Iain Stewart.

Figure 6.3 'Temple of Serapis', Pozzuoli, Naples. Note indicators of former shorelines on pillars. Image credit: Shutterstock/lauradibi

anticipate that all coastal areas affected by vertical tectonic movements have little value to scientists attempting to reconstruct past changes in sea level. For example, a Holocene sea level curve for eastern Japan will provide limited information on past glacio-eustatic changes, but will be enormously valuable in the reconstruction of former earthquake activity. In some coastal areas, past rates of land uplift have been high. One such example is that of the uplifted coral terraces of the Huon Peninsula in Papua and New Guinea, studied for many years by John Chappell, where the reconstructed long-term uplift rate is in the order of three metres per thousand years.

Whereas long-term average rates of tectonic uplift have been used in this way to calibrate sea level curves, one also needs to be aware that irregular vertical land movements can also take place at much shorter timescales, ranging from centuries to the near-instantaneous effects associated with earthquakes. One of the most famous examples of short-term vertical tectonic movements characterised by both uplift and subsidence derives from Charles Lyell's studies of the Roman Macellum or 'Temple of Serapis' in Pozzuoli, near Naples in 1828 (Fig. 6.3). Lyell observed a line marked by marine *Lithophaga* bivalve molluscs on several pillars, and concluded that this was indicative of a former shoreline. He proposed that this coastal area had been submerged for a lengthy period of time after the Roman era. Thereafter, it was subject to crustal uplift that caused the former shoreline position to be raised to just under 3m above present. The shoreline has been subject to further tectonic uplift since

Figure 6.4 Emerged shoreline notches in an aseismic region, South Gigante Island, Philippines. Image credit: Max Engel.

Lyell's time, these processes most probably linked to dynamic changes within adjacent magma chambers in the Naples area.

Consider Figure 6.4, which shows two uplifted shoreline notches in a part of the Philippines. How are we to tell whether these features have been uplifted to their present position as a result of past earthquakes? Conversely, is it the case that the area is completely aseismic (earthquake-free) and that the notches were produced by 'normal' processes of bio-erosion during periods of time in the past when relative sea level was higher than present? (Compare with Figure 6.2.)

These questions illustrate and exemplify the considerable difficulties faced when one tries to separate the effects of past vertical tectonic movements from a specific sea level curve. Imagine a hypothetical scenario where a contemporary shoreline feature was uplifted by +5m during an earthquake. Thereafter the coastal area experienced post-seismic subsidence of –2m over the following 2000 years. This would mean that after 2000 years the feature could be observed in the field at +3m. How then are we to know that this, now raised, feature was originally uplifted to + 5 m? The latter conclusion can only be drawn if there are other pieces of field evidence that demonstrate an initial co-seismic uplift of +5m (Figs 6.2 and 6.4).

From the above, it is clear that coastlines in areas affected by co-seismic, post-seismic and inter-seismic movements will display chronological sequences of long-term sea level changes that are incredibly complex, showing relative sea level lowering in some areas and relative sea level rise in others and, in some cases, patterns of irregular vertical 'seesaw' movements involving phases of both uplift and subsidence.

Geoidal eustasy

We have to go back over a century (1888) to discover RS Woodward of the US Geological Survey describing how deformation of the ocean surface takes place adjacent to large masses of ice, and also how the melting of ice sheets is accompanied by a geoidal lowering of the adjacent ocean surface. In the decades that followed, a number of eminent geologists and geophysicists, including Daly, Bloom, Walcott, Chappell, Cathles, Fjeldskaar and Andrews all advanced our understanding of this issue in numerous ways. In particular, in 1976, James Clark wrote an influential short paper that described the role of ice–water gravitational attraction in determining patterns of relative sea level change lowering that accompanied the (partial) melting of the Greenland ice sheet at the end of the last ice age. During the following years, Clark together with co-authors William Farrell and Dick Peltier wrote a series of papers that were to change completely the way that we understand past changes in relative sea level. Of the papers that were written, one figure produced by Willy Fjeldskaar in 1989 summarised the new thinking concisely (Fig. 6.5). Fjeldskaar used a very simple model to demonstrate how the melting of the last ice sheet across Scandinavia led to a slackening of the gravitational attraction of ocean water to the ice sheet. He showed how a theoretical melting of this ice c.15,000 years ago

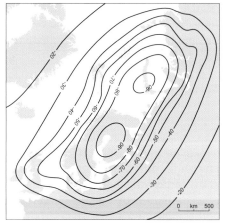

Figure 6.5 Patterns of geoidal sea surface lowering (m) associated with an instantaneous melting of the last Fennoscandian ice sheet. Image credit: adapted from W. Fjeldskaar.

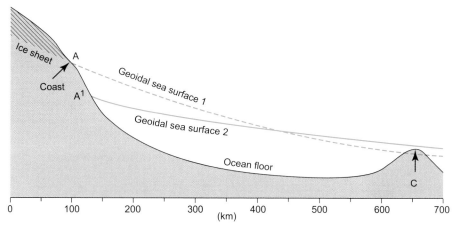

Figure 6.6 Schematic illustration of geoidal sea surface changes associated with the melting of an ice sheet. Island C is located in the open ocean c.700km from coastline (A and A1). Melting of the ice causes gravitational relaxation between ocean and ice sheet. This results in a geoidal sea surface lowering from A to A1. Island C experiences submergence (note that the effects of glacio-eustatic changes are excluded from this illustration). Not to scale.

would have led to a theoretical lowering of the sea surface as far away as Scotland by c.–50m. Of course such a lowering would have been countered by a glacio-eustatic rise in sea level. This very simple piece of research illustrates the importance of geoidal sea surface changes in determining the actual patterns of sea level change that happen at the coast when an ice sheet is subject to melting (Fig. 6.6).

Isostasy

The 'stasis' or equilibrium of the Earth's crust in response to its surface loads of water and ice is the essence of the concept of isostasy. Since the nature of these loads is subject to constant change, isostatic deformation of the lithosphere is also taking place constantly, albeit at extremely small rates. During the last glacial maximum, however, the distribution

of ice and water loads was quite different. When the ice started to melt it set in train a chain of crustal processes that remarkably still continue today, c.20,000 years later! There are two main isostatic processes relevant to our understanding of sea level change. The first is the process of glacio-isostasy and refers to crustal changes associated with the growth and decay of ice sheets. The second is the process of hydro-isostasy, which is associated with deformation of the oceanic lithosphere in response to changing loads of ocean water.

Glacial isostasy

In 1865 Thomas Jamieson, a Scottish geologist, first described the process of glacial isostasy. Although not specifically using the word 'isostasy' in his paper he noted that 'It is worthy of remark that in Scandinavia and North

America, as well as in Scotland, we have evidence of a depression of the land following close upon the presence of the great ice-covering.' In a paradigm-shifting statement he added that 'it has occurred to me that the enormous weight of ice thrown upon the land may have had something to do with this depression (and that) the melting of the ice would account for the rising of the land'. It followed also that the greatest amount of land rebound would occur in areas where the ice was formerly thickest, with the amount of glacio-isostatic rebound decreasing towards the margins of any former ice sheet (Fig. 6.7).

In the decades that followed Jamieson's idea, scientists in Europe and North America observed that individual shorelines in areas affected by glacio-isostatic rebound decline in altitude in directions away from

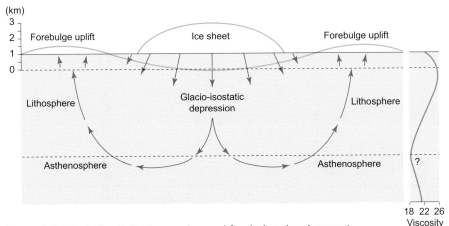

Figure 6.7 Glacio-isostatic depression and forebulge development associated with ice-sheet growth. Note the role played by a low-viscosity asthenosphere. Profile on right shows an idealised viscosity profile through the lithosphere and asthenosphere.

Figure 6.8 Raised shorelines of NE Islay, Scotland, produced as a result of interaction of glacio-isostatic rebound and glacio-eustatic relative sea level rise. Image credit: A.G. Dawson.

the centre of glacio-isostatic uplift. Some of these shorelines that exhibit evidence of regional tilting due to differential glacio-isostatic rebound represent shorelines of ice-dammed lakes that were ponded against former ice-sheet margins (e.g. glacial lake Agassiz in North America) while others were formed by marine processes. The oldest such shorelines (i.e. those that formed during the earliest stages of ice sheet decay) generally experienced the greatest amounts of glacio-isostatic rebound. Not surprisingly, the absolute amount of glacio-isostatic rebound varies according to the dimensions of the ice sheet that formerly covered the respective landscape. Thus the melting of the last (Laurentide) ice sheet across North America was associated with much greater amounts of rebound when compared with, for example, the last Scottish ice sheet (Fig. 6.8).

The growth of any large ice sheet over thousands of years is accompanied by deformation of the underlying lithosphere and asthenosphere. These processes are not just confined to the area directly beneath the ice, but extend laterally several hundred kilometres outwards beyond the ice margin. After an ice sheet has completely melted, the process of crustal rebound continues for many thousands of years. Indeed, this timelag is so large that extensive areas of the Earth's surface in formerly glaciated landscapes continue to undergo this rebound at the present day, albeit in small amounts. In certain areas where the ice cover was thickest (e.g. northern Sweden and Hudson Bay, Canada) the present rates of land uplift are as much as $c.10\,\mathrm{mm\,yr^{-1}}$ (Fig. 6.9).

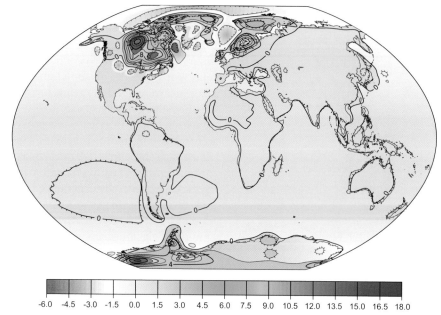

-6.0 -4.5 -3.0 -1.5 0.0 1.5 3.0 4.5 6.0 7.5 9.0 10.5 12.0 13.5 15.0 16.5 18.0

Figure 6.9 Patterns and rates of lithospheric rebound (mmyr⁻¹). Note the distribution of regions of forebulge collapse around margins of former ice sheets. Image credit: NASA, JPL, Eric Ivens.

Glacio-isostatic processes associated with ice sheet growth and decay involve both land uplift and subsidence. In simple terms, if a landscape is loaded by an ice sheet (with a density of slightly less than 1 (g/cm³)) on an underlying lithosphere (with an average density of c.2.5–3.0), the land surface will be lowered by an amount inversely proportional to the density ratio between ice and rock. The vertical deformation of the underlying lithosphere is, in turn, dependent on displacement of sub-crustal magma in the asthenosphere. As the ice sheet builds up, the movement of subcrustal material in the asthenosphere is outwards and towards areas beyond the ice sheet margin. In areas peripheral to the developing ice sheet, the lithosphere is pushed upwards, causing a forebulge (Fig. 6.7). One of the first to describe these processes was Fridtjof Nansen in 1922. He described 'Supposing the ice-cap begins to be formed in the central area of an extensive region like Fenno-Scandia, the load of the ice cap will press the crust down in this central area, and in a zone surrounding it the crust will be pressed up and will there form a kind of concentric wave … (that) … will gradually extend outwards, and will be flattened down as it becomes wider and wider.'

When an ice sheet melts, these processes are reversed. Subcrustal material is gradually returned to the asthenosphere regions underneath the formerly ice-covered areas, leading to a rebound of the lithosphere. Much of this glacio-isostatic rebound takes place while the landscape is still covered by ice, since it is the thinning of the ice sheet that stimulates the vertical rebound. By contrast, the area of crustal forebulge starts to subside as soon as sub-crustal material starts to migrate back towards the areas where glacio-isostatic depression had previously taken place.

The processes of glacio-isostatic deformation depend on the viscosity of the asthenosphere. A high-viscosity asthenosphere will respond relatively slowly to loading of the lithosphere by ice. By contrast, a low-viscosity asthenosphere will respond much more quickly to ice loading and unloading. The problem, as we have seen, is that scientists are uncertain of the actual viscosity of the asthenosphere, and it becomes quite difficult to quantify and interpret patterns of glacio-isostatic deformation associated with any former ice sheet. This problem keeps recurring in various forms when attempts are made to interpret present-day rates of relative sea level changes in areas formerly covered by ice.

Hydro-isostasy

The 'stasis' or balance referred to earlier applies also to loads exerted by ocean water onto the sea floor. In this way the melting of ice sheets results in increased ocean mass, which results in deformation of the underlying lithosphere and this, in turn, leads to displacement of material within the asthenosphere. Two consequences arise from these processes. First, the seabed sags downwards due to increased ocean mass, resulting in a different shape of ocean basin within which the seawater is

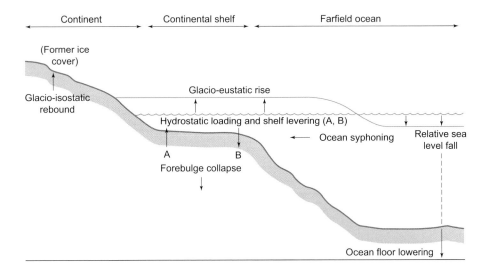

Figure 6.10 Schematic illustration of key relationships between long-term crustal movements and changes in ocean mass. The sketch also includes a depiction of glacio-eustatic change that includes a schematic inflection for farfield ocean areas to highlight the effect of ocean-floor lowering (sagging).

contained. Second, the subcrustal material within the asthenosphere moves outwards and away from the zone of lithosphere displacement (Fig. 6.10).

These two processes (ocean mass increase and seafloor deformation) tend to occur out of phase with each other. The increases in ocean mass occur relatively rapidly in response to global ice melting, whereas deformation of the sea floor takes place at a much slower rate. During the periods of the most rapid deglaciation that accompanied the decay of the last great ice sheets, average sea level may have risen at rates in the order of metres per century (see chapter 9). At the same time the deformation of the ocean floor took place much more slowly. One of the most important factors that influenced the pattern and rates of such change was the variable thickness of the oceanic

lithosphere. In areas where the lithosphere was relatively thick the degree of warping of the sea floor was relatively small. In other areas, for example near hotspots where the oceanic lithosphere was relatively thin, the deformation was much more pronounced.

Along the margin of continents the hydro-isostatic changes are more complex than in open ocean areas. One key factor relates to the shallower water depths. Since the melting of the last great ice sheets resulted in an average rise of sea level in the order of +120m, the effect of this additional water load across, for example, a continental margin would be different from a similar additional point load of water added in the open ocean. These processes resulted in the tilting of many continental margins, with the outer areas effectively being pulled down in parallel with

the subsiding ocean floor. In this way, continental margins tend to experience flexure that takes place in conjunction with the addition of increased volumes of ocean water. Where such continental margins coincide with trailing edges of continental plates and wide continental shelves, this flexure is most evident. As long as the ocean floor continues to deform with increased water loading, the movement of subcrustal material within the asthenosphere will generally progress in a direction towards and beneath the continental lithosphere and cause parts of the inner continental shelf to be levered upwards (Fig. 6.10).

Such hydro-isostatic changes are of paramount importance to understanding the ways in which large ocean areas respond to increased loads of seawater. The timelag between the addition of ocean water load and the response of the oceanic lithosphere means that the latter continues to sag downwards thousands of years after the ocean volume increase from ice melt has ended. For the Pacific Ocean, the progressive deepening of the ocean basin throughout much of the Holocene has led to coastal emergence and the occurrence of raised shoreline features. At first sight such features might be interpreted as evidence of a formerly warmer global climate and higher sea levels – in fact they demonstrate no such thing!

Steric changes

Steric changes refer to changes in the *sea surface* rather than to sea level. Such changes refer, therefore, to changes in ocean volume where there is no change in ocean mass. The most well-known steric processes include sea surface changes arising from short-term changes in air pressure and longer-term changes in ocean density and salinity. The passage of cyclones across an ocean area also results in a general increase in sea surface elevation. In general, a cyclone with a central air pressure of 950mb is capable of raising the regional sea surface by *c*.0.6m. This pressure-induced change is popularly known as the inverse barometer effect. The increase in sea surface elevation is short-lasting, however, and will disappear with the passage of the storm.

Changes in ocean temperature do not by themselves cause changes in the sea surface. More specifically, changes in ocean temperature lead to changes in ocean water density, and these density changes, in turn, cause the regional sea surface to change

in elevation. Such changes result in changes in ocean volume and in the buoyancy of ocean water but not ocean mass, and these changes may be long-lasting. Several processes may contribute to changes in the volume of a fixed mass of water. Perhaps the most widely discussed is the process of thermal expansion of ocean water (or contraction) caused by changes in ocean density and sea temperature (Fig. 6.11). Empirical measurements of time-series of ocean temperature changes have long been known to indicate that past changes in regional sea surfaces have taken place as a result of ocean density changes. Heating of the ocean surface has become an issue of considerable importance in recent years as a result of the debate on the nature of climate

change. A popular citation is that as much heat is stored within the first three metres of the world's oceans as is contained within the entire atmosphere. Even if this observation is remotely close to being true, it serves to highlight the fundamental role that the oceans play as a heat source.

The issue of thermal expansion of ocean water is of fundamental importance to the current debate on the nature of recent sea level rise and how the world's oceans will respond in the future to increased air and ocean temperatures. With virtually every numerical model of future climate change predicting increases in atmospheric CO_2, it follows that all of the models also predict rising ocean temperatures, reduced ocean water density, and therefore increased thermal expansion of ocean water.

Steric changes in the sea surface can also occur as a result of more widespread disturbances in weather and climate. This is well illustrated for the Pacific Ocean, where the occurrence of an El Niño event will heat ocean water and, through alterations in regional ocean circulation, will also 'push' ocean water towards the Pacific coastlines of the United States, Central and South America. The effect of each major El Niño event over the last century is recorded on tide gauges in these areas. For example, the San Francisco tide gauge record that extends as a continuous record from 1897 to the present shows the largest El Niño events (1982–83 and 1997–98) very clearly where they each resulted in average sea surface increase routinely in excess of 30cm above mean sea level for several months (Fig. 6.12).

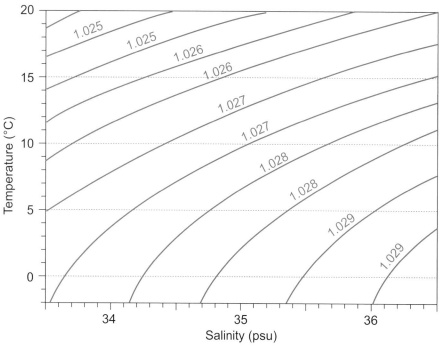

Figure 6.11 Numerical relationship between ocean temperature, density and salinity.

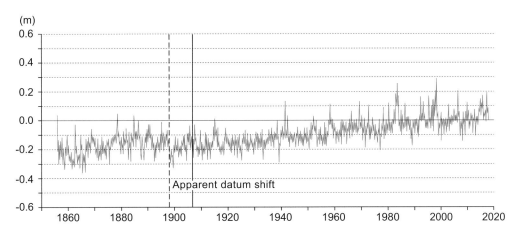

Figure 6.12 Tide gauge record for San Francisco. The later part of the graph indicates a mean sea level rise for the 20th century of +1.94mmyr^{-1}. Image credit: NOAA.

Whereas at short timescales (i.e. last century) the thermal expansion of ocean water is a key factor contributing to recent rates of sea level change, it is not generally considered at longer timescales, for example during the Holocene. This is because rates of sea level change during the early Holocene were so much greater than present that any steric effect due to thermal expansion would be lost as 'noise'. At these much longer timescales, geoidal-eustatic changes leading to changes in the distribution of ocean water would have been considerably more important.

7 Ancient and modern ice sheets and relative sea level changes

Introduction

Ever since Shackleton's original work, marine oxygen isotope curves have been produced for much of the Earth recent history (Fig. 7.1). The curves that cover the most recent intervals of geological time are the most detailed (Fig. 7.2). The time intervals are subdivided into marine oxygen isotope stages and substages. Stage numbers that are even correspond to major ice ages with odd-numbered stages as interglacials. Isotope substages (3a, 5e, etc.) represent subdivisions of stages where there are good reasons, on grounds of palaeoclimate, to subdivide individual stages. Thus, whereas the last glacial period (isotope stage 2) has no substages associated with it, isotope stage 5 is subdivided into five substages. Foremost amongst these is isotope substage 5e, which is generally considered to be equivalent to the warmest period of the last interglacial. Numerous published marine oxygen isotope curves exist, and many extend much further back in time and cover numerous glacial (2, 4, 6, 8, 10, 12, etc.) and interglacial (5, 7, 9, 11, 13, etc.) stages. Similarly, the lettering of isotope substages refers to periods of relative warmth and cold. Thus within isotope stage 5, substages 5b and 5d represent relatively cooler conditions, while substages 5a and 5c correspond

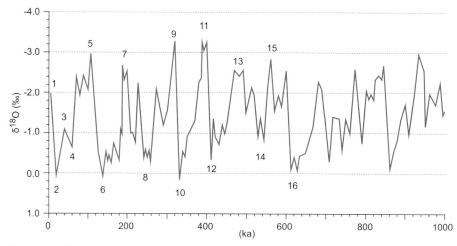

Figure 7.1 Marine oxygen isotope variations over last *c*.1Myr. Image credit: adapted from J.J. Lowe and M.J.C. Walker (2014).

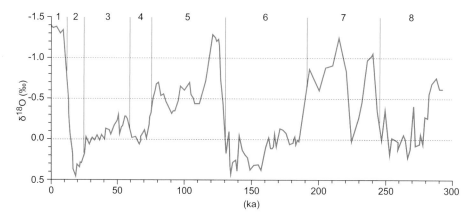

Figure 7.2 Marine oxygen isotope variations over last *c*.300,000yr. Image credit: adapted from J.J. Lowe and M.J.C. Walker (2014).

to periods of relative warmth.

It should be remembered that the time resolution of specific ocean sediment cores is dependent on past ocean floor sediment accumulation rates, as well as being limited by the amount of sediment bioturbation (burrowing within surface sediment by marine organisms) that has taken place. In general, ocean floor sediment cores taken from locations where rates of past sedimentation have been high tend to provide more precise information on past environmental changes. Significant amounts of bioturbation can often lead to significant losses of detail in ocean cores. For these reasons, the most detailed information on past oxygen isotope variations are from cores covering the most recent period of geological time and from areas of relatively rapid ocean floor sedimentation.

The marine oxygen isotope stratigraphy of most ocean floor sediment cores generally includes isotope substage 5e, centred on c.125,000 years ago, which corresponds to the last period of maximum interglacial warmth. In terms of the 'old' nomenclature of glacial and interglacial periods, marine isotope substage 5e corresponds with the Eemian interglacial in Europe, the Ipswichian interglacial in the UK and the Sangamon interglacial in North America. Similarly, isotope stages 2–4 are grouped together to represent the Weichselian glaciation in Europe, the Devensian glaciation in the UK and the Wisconsin glaciation in North America. Differences in stage numbering also exist. For example, in Russia the established terminology has been to define the Valdai glacial period as beginning in isotope substage 5d and having culminated during stage 2 (Table 3.2).

In terms of global ocean volume changes (and hence sea level changes), some important patterns of change are apparent through inspection of the marine oxygen isotope curve (Fig. 7.2). For example, it should be noted that during the isotope stage 5/4 transition there is a very large ^{18}O change equivalent to an average lowering of global sea level by c.50m. Best estimates place the onset of this change around c.70–80,000yr BP. Clearly the widespread growth of ice sheets took place around this time, but crucially we do not know where the principal areas of ice accumulation were located.

For some parts of the world the palaeo-environmental reconstructions for isotope substages 5a–d are quite detailed. Thus in North America, scientists have suggested that this period, known as the Eowisconsin, was characterised by two glacial advances that may have taken place across the Canadian Arctic and possibly as far south as the St Lawrence Lowlands. Similar patterns of change are thought to have occurred in Scandinavia. Here, studies of cave sediments along the western coastline of Norway point to major growth of ice sheets during isotope substages 5b and 5d as well as during the stage 5/4 transition (the latter known as the Karmoy Stadial). After isotope stage 4, the marine isotope record shows a perceptible depletion in ^{18}O that would appear to be indicative of a significant amount of ice sheet melting worldwide. As mentioned, it is not possible to establish at the present time where this loss of ice took place. The most plausible explanation is that the melting took place across several continents rather than in one single region.

As can be gathered from above, scientists have very limited information about the nature and scale of the climate changes that took place following the end of the last interglacial (substage 5e). Environmental reconstructions for this time period depict a picture of ice sheets of contrasting dimensions waxing and waning at different rates and by different amounts across different parts of the world. Since the ice sheets would have experienced different glacio-isostatic histories, all that can be inferred in terms of sea level reconstructions for this period of time is that almost nothing is known of the trends in relative sea level for any of these areas.

Global ice volumes, ocean volumes and sea level equivalents

The measurement of the change in the isotopic composition of seawater for the last glacial maximum is associated with values between 1.2 and 1.7 ^{18}O per millilitre in relation to the modern standard reference value – equivalent to an absolute reduction in ocean volume for isotope stage 2 of between 47 and 65 million km^3. This range of values is in broad (but not precise) agreement with estimates based on mapping and dating the former extent of ice cover across the Earth. The latter is particularly difficult to accomplish for two key reasons. First, the margins of many of the ice sheets that existed during the last glacial maximum are located below present sea level on continental shelves. Many are not mapped in detail (e.g. offshore Greenland and

Antarctica) and the ages of only a very few submerged ice-marginal moraines are dated with any accuracy. Second, it is difficult to determine the thicknesses of the ice sheets that covered continental landmasses during the last glacial maximum. Scientists have generally tried to address this problem by the use of idealised ice sheet profiles generated by mathematical models. Both of these problems come together when attempting to estimate the dimensions of the Antarctic and Greenland sheets for the last glacial maximum. Of these two ice sheets, the ice cover across Antarctica is by far the bigger. One popular estimate is that the volume of Antarctic ice during the last glacial maximum was c.38 million km^3. If one subtracts from this the present volume of ice in Antarctica, we are left with an additional c.10 million km^3 of ice that could contribute to average sea level lowering worldwide.

Estimated volumes of ice in respect of Greenland are considerably smaller. A popular estimate of the additional ice volume that existed during the last glacial maximum in Greenland is c.2.5 million km^3. In addition to these changes the two other key ice sheets that existed during the last glacial maximum were in Scandinavia and Russia (the Fennoscandian ice sheet often referred to as the Eurasian ice sheet) as well as in North America (known as the Laurentide ice sheet) (Fig. 7.3). In respect of the former, best estimates suggest somewhere between 8 and 13 million km^3 of ice contributing to average global sea level lowering. However, the Laurentide ice sheet that developed in North America during the last glacial maximum contributed approximately 30–34 million km^3 of ice

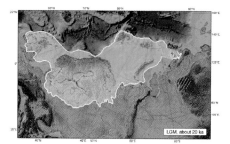

Figure 7.3 Reconstruction of dimensions of Eurasian ice sheet during the last glacial maximum. Image credit: J. Mangerud and J.-I. Svendsen.

to sea level lowering. When all of these ice volumes are added together, they point to a sea level lowering equivalent volume that lies somewhere between c.97 and 117m depending on the various sources of error described above, and excluding the effect of a greater cover of valley glaciers and ice caps at this time. Notably the maximum ^{18}O enrichment at the last glacial maximum measured from ocean sediment cores is broadly in agreement with this estimate range for sea level lowering. In terms of sea level equivalent, the values listed above point to the Laurentide ice sheet during the last glacial maximum having been by far the most significant contributor to

an average sea level lowering worldwide (c.34m). That said, caution needs to be exercised with these calculations, since hydro-isostatic deformation of ocean floors may have altered the capacity of ocean basins to accommodate a much reduced volume of ocean water.

Remarkable sea levels during marine isotope substage 5e

One of the most striking aspects of the marine oxygen isotope curve for the last c.130,000 years is that during isotope substage 5e the ^{18}O values from the foraminifera data are depleted to such an extent that the isotopic deviation reaches values that are greater than the reference standard for the present. This means that during the peak of substage 5e there was less ice stored in continental ice sheets and glaciers than there is at present. This observation is consistent with the geological field evidence of raised shorelines for different areas around the world (Fig. 7.4). For example, in Portugal average sea level is considered to have been between 1 and 5m higher than at present. In eastern North America, a raised shoreline belonging

Figure 7.4 Emerged reef platform, Spelonk lighthouse, NE Bonaire, Dutch Caribbean dated to last interglacial (substage 5e). Image credit: Max Engel.

to the last (Sangamon) interglacial is a prominent feature several metres above present. One has to be careful in choosing examples to illustrate the former occurrence of such high sea levels, since it is possible to observe raised shorelines of this general age at these typical altitudes, or indeed much higher altitudes in areas of tectonic uplift (e.g. the Pacific coastline of Chile). That said, there is sufficient field evidence to point to the occurrence of substage 5e sea levels several metres higher than present for areas where tectonic uplift has been negligible over the last *c.*130,000 years.

The most important observation to be drawn from this information, and something always to be remembered in current debates on climate change and sea level rise, is that the scale of ice melt during the last interglacial period was sufficiently high to enable sea level around the world to reach altitudes greater than present – without the agency of anthropogenic-induced warming! Indeed, in northern Scandinavia sea level was sufficiently high to enable the creation of a marine connection between the Baltic Sea and the Arctic Ocean. Scientists have struggled to find a satisfactory explanation for such high sea levels. After all, analysis of Greenland and Antarctic ice core data points to the fact that atmospheric concentrations of carbon dioxide during the last interglacial reached no higher than *c.*280–290 parts per million by volume (ppmv). In other words, the global warming that took place during substage 5e cannot be explained by a dramatic increase in greenhouse gas concentrations. Nor are average global air temperatures considered to have

been significantly higher than present at that time. In summary, the two principal candidate processes (higher greenhouse gas concentrations and higher global air temperatures) that are used at present to account for rising sea levels cannot be used to explain the higher glacio-eustatic sea levels that appear to have existed during substage 5e. Instead, two other theories have been advocated.

Over 40 years ago, John Hollin and John Mercer suggested that part of the West Antarctic ice sheet may have disintegrated during marine isotope substage 5e. They maintained that sufficient ice could have melted during substage 5e to have raised sea levels worldwide by up to 5m. This idea has been discussed many times since Hollins' and Mercer's remarkable suggestion (Fig. 7.5). But in order to understand the basis for this reasoning, it is first necessary to consider the nature of the ice cover that existed in Antarctica during isotope substage 5e.

The East and West Antarctic ice sheets

It has long been known that if all of the ice across Antarctica were to melt, sea level would rise somewhere in the order of 60m, sufficient to cause global catastrophe. Indeed, maps have been published in popular magazines to show what the scale of marine inundation across the world would look like if such a rise in sea level were to take place in the future. The reality is, however, more complex. According to 'popular' awareness, the ice cover across Antarctica is a single entity – this is simply not true. Two major ice sheets presently exist in Antarctica – the East and West Antarctic ice

Figure 7.5 NASA reconstruction of areas of Antarctica that have experienced the greatest surface warming between 1957 and 2006. The dark red areas depict regions where there has been an overall increase of 0.5°C since 1957, whereas the light red areas illustrate zones of intermediate overall warming (white denotes no overall change). Note that the instability or otherwise of the West Antarctic ice sheet is strongly influenced by migrations of ice sheet grounding lines. Image credit: NASA.

sheets. The bigger by far is the East Antarctic ice sheet. It has a present ice volume in the order of 23 million km³ of ice, but it is a cold-based ice sheet, meaning that the vast majority of the base of the ice is frozen to the underlying bedrock. This means that the ice sheet is essentially stable, and there is no known mechanism that would enable it to melt catastrophically. Furthermore, the ice sheet covers such a huge area that it generates its own permanent anticyclone as a result of the ice surface continually cooling the overlying air and causing the air to sink. Thus, very little snow is delivered annually to the central areas of the ice sheet except as a result of wind activity. Locally, particularly around the margins of the ice sheet, annual changes in mass balance are more marked. For example, increases or decreases in snow precipitation may cause particular ice masses close to the margin of the ice sheet to increase or decrease in size. Also rates of calving of tidewater outlet glaciers may change over time. But to all intents and purposes, the East Antarctic ice sheet does not figure prominently in scientifically based discussions of present day sea level changes.

This is not the case in terms of longer-term glacial–interglacial changes, in particular in respect of the growth and expansion of the East Antarctic ice sheet during times of global climate cooling. For example, in respect of the growth and expansion of ice cover during the last glacial maximum there is general agreement that the East Antarctic ice sheet increased in size significantly, extending further out onto its continental shelf areas. The additional ice is estimated as between c.2.5 and 7.2 million km³, equivalent to between 6 and 18m of global sea level equivalent.

The West Antarctic ice sheet is quite different. Whereas the East Antarctic ice sheet is located across the Antarctic continent, the West Antarctic ice sheet is located over a mountain chain that represents the southern extension of the Andean mountain chain (Fig. 7.5). The most important consequence of this is that much of the glacio-isostatically depressed land beneath the ice sheet is presently located below sea level. This means that seawater has the potential to destabilise ice sheet grounding lines (Fig. 7.6). The specific areas where this may occur are largely unknown, although the Amundsen Sea area is considered to be particularly susceptible to

grounding line instability (see chapter 11). In this way, accelerated discharge of ice through outlet glaciers emanating from the interior ice sheet could lead to changes in the mass balance of the ice sheet and its eventual collapse.

Such a scenario was described by Hollins and Mercer in respect of isotope substage 5e, and this view has been reiterated by many scientists since then. In essence, whereas ice across East Antarctica is relatively stable, cold-based, below pressure melting point and unlikely to change dramatically due to global warming, the ice in West Antarctica is potentially unstable and susceptible to collapse under a warm interglacial climate. Since the volume of ice in West Antarctica (c.2.2 million km³) equates to roughly 5.5m of global

sea level, the possible disintegration of much of this ice during isotope sub-stage 5e has always made it a likely candidate to account for the higher sea levels worldwide at this time.

The Greenland ice sheet during substage 5e

In contrast to the West Antarctic ice sheet disintegration hypothesis, others have argued that the higher than present sea levels during isotope substage 5e can be explained by the partial melting of the Greenland ice sheet. Over 20 years ago, Nils Reeh put forward the idea that the southern part of the Greenland ice sheet may have disappeared during this time interval. His view was partly based on ice sheet modelling and also on the results of

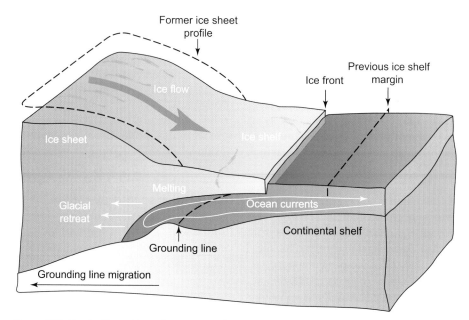

Figure 7.6 Illustration of ice sheet grounding line migration. The issue of key importance is whether or not marine water can penetrate landward of the grounding line, causing the ice sheet to become destabilised.

ice core drilling that failed to indicate the presence of any ice at drilling sites across southern Greenland that was as old as substage 5e. Reeh suggested that during isotope substage 5e the Greenland ice sheet split into a main ice sheet that covered central and northern Greenland, but that much of the southern part of today's ice sheet had melted, leaving a smaller residual ice cap located over the south-eastern highlands. Estimates of the amount of ice loss are uncertain, and as a result it has always proved near impossible to estimate an accurate value of the amount of sea level rise equivalent to the amount of ice that may have melted from Greenland at this time.

Lest we forget, the marine oxygen isotope chronology points to the same conclusion of higher sea levels than present during isotope substage 5e. Whether the principal ice melt originated in Greenland or West Antarctica is still a matter of debate. However, the outcome of this discussion is crucially important, since it has a direct bearing on the current predictions of future changes in relative sea level (see chapter 11).

Episodes of rapid sea level rise during the last *c*.150,000 years

Inspection of the marine oxygen isotope record for isotope stage 6 through to the present point to three distinct periods of time when sea level rose rapidly around the world (Fig. 7.2). The first of these appears to have been during the transition between isotope stages 6 and 5e, the second is during stage 3, while the third corresponds to the transition between stages 2 and 1. Of these the last is the best known, since it corresponds with the melting of the last great ice sheets on Earth and the transition into the present Holocene interglacial. A similar pattern of sea level rise appears to have taken place during the melting of ice sheets worldwide at the close of stage 6. The oxygen isotope record points to the amount of glacio-eustatic fall during stage 6 having been of the same order as during isotope stage 2. This observation points to the likelihood that the scale of ice sheet glaciation was broadly similar during stages 2 and 6. If this indeed was the case, one may reasonably infer that the amounts of glacio-eustatic fall on both occasions were also broadly equivalent.

The possibility of a major episode of sea level rise during isotope stage 3 is also a feature of the marine oxygen isotope record. This particular time interval may be significant since, apart from the present Holocene interglacial, it is the only other period of time since isotope stage 6 that the trend of relative sea level was upwards and that the amount of glacio-eustatic sea level rise was large (in the order of *c*.50m). That said, we know very little indeed about patterns of environmental change that may have taken place around this period of geological time. Unlike the palaeo-environmental changes that took place during the transitions between stages 2 and 1 and also between 6 and 5e, this apparent rise in sea level took place at a time when the Earth's climate was already experiencing cold stadial conditions, yet neither as cold as during a period of general glaciation nor as warm as during an interglacial.

Late Pleistocene sea level changes and the Huon Peninsula uplifted reefs

The uplifted reef sequences of the Huon Peninsula, Papua and New Guinea also provide valuable information. John Chappell and his team identified within the reef shoreline 'staircase' an uplifted terrace dated to marine oxygen isotope stage 5e (*c*.125,000yr BP) that was compared with other coastal features of the same approximate age elsewhere (e.g. Barbados). Individual terraces above and below this level were also dated by ^{230}Th/^{234}U to provide a chronology of terrace formation that extended back nearly 150,000 years. When the Huon terrace altitude/age data is superimposed on the marine oxygen isotope curve and adjusted for long-term vertical tectonic movements, the fit is nearly perfect (Fig. 3.3). From this, one can conclude two possibilities. The first is that the ages and altitudes of the Huon terraces, when corrected for past uplift, provide an independent verification of the marine oxygen isotope record for the Late Pleistocene. Equally, one can interpret the data in reverse – that the marine oxygen isotope record provides an independent verification of the glacio-eustatic curve of relative sea level derived from the Huon terrace data.

8 Relative sea level change during the last glacial maximum

What is so important about knowing the positions of relative sea level during the last glacial maximum, and what has it to do with the current debate on sea level rise? The answers to both of these questions lie in the fact that the last ice age effectively 'sets the stage' for present day patterns of relative sea level change on Earth. It may seem remarkable, but the Earth's lithosphere is still readjusting to patterns of glacio-isostatic deformation that began over 20,000 years ago. As mentioned earlier, the ice sheets that developed at this time were so large that they led to crustal depression of the underlying lithosphere on a grand scale. Not only that, but vast areas of terrain located beyond the margins of these huge ice sheets were also subject to crustal deformation. The vertical uplift of these areas of crustal forebulge represented a response to the crustal depression that took place underneath the ice sheets. As the great ice sheets increased in size, continental margins experienced flexure, while ocean floor areas located distant from the ice sheets adjusted in complex ways to decreased loading by ocean water.

Our understanding of ice age sea levels is further complicated by additional effects. For example, the ice sheets altered the Earth's gravity field as a result of large-scale redistribution of mass across the Earth's surface. The Earth's geoidal sea surface changed radically. One of the most significant differences was that the geoidal sea surface was drawn upwards towards the margins of the respective ice sheets (this process also happens today around Antarctica and Greenland, where the geoidal sea/ocean surface experiences gravitational attraction towards the ice sheet) (Fig. 8.1). The growth of large ice sheets across the northern hemisphere was also associated with changes in the Earth's radius and its rate of rotation, as well as the 'drifting' of the Earth's axis of rotation.

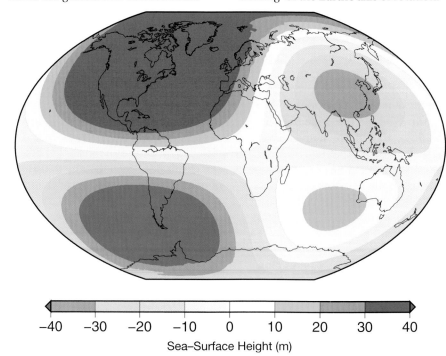

Sea–Surface Height (m)

Figure 8.1 The gravitational pull on ocean water by the great northern hemisphere ice sheets (not shown) during the last glacial maximum was immense. Modelling research by Glenn Milne shows that c.20,000 years ago relative sea surface height compared to the present day was considerably higher in the northern hemisphere than across the southern hemisphere. This reconstruction includes the effects of changes in Earth rotation properties and uses Dick Peltier's ICE-5G ice model and the associated VM2 viscosity model. Image credit: Glenn Milne.

The reason why ice-age sea levels are important to our understanding of present day relative sea level change is because the Earth continues to experience the consequences of the last ice age that ended *c*.20,000 years ago. The quantification of this glacio-isostatic adjustment is principally derived from numerical modelling of relative sea level change; indeed, a numerical value for this glacio-isostatic adjustment is included within the most recent assessments of current rates of relative sea level change by the Intergovernmental Panel on Climate Change (IPCC) (see chapter 11). Clearly, it is important to be able to determine whether or not modern processes of glacio-isostatic adjustment serve to increase current rates of sea level rise beyond what they would otherwise be as a result of climate change processes. Or is glacio-isostatic adjustment currently functioning in an opposite way, such that it counteracts the trend of rising sea levels? Furthermore, what is the size of the glacio-isostatic adjustment component, and across what areas of the Earth does it have its greatest effect?

For the field scientist studying patterns of former relative sea level change for a particular area, the above questions are impossible to answer. Numerical modelling makes it possible to provide some answers, but the modelling results are only as accurate as the quality of the data used in the running of the models. Central to the accuracy of the models are answers to the following questions:

◆ Can an age be assigned to the last glacial maximum after which ice sheets started to melt?

◆ Can it be assumed that all of the ice

sheets during the last glacial maximum were in glacio-isostatic equilibrium?

◆ What was the geographical extent of these ice sheets during the last glacial maximum and what were their dimensions – what volume of ice was stored within them?

◆ Can a similar calculation be made for all ice caps and valley glaciers for the last glacial maximum?

◆ Did all the former ice sheets reach their maximum extent at the same time?

◆ How easy (or difficult) is it to reconstruct palaeo-coastlines for land bridges that existed during the last glacial maximum?

Can an age be assigned to the culmination of the last glacial maximum after which ice sheets started to melt?

The first wide-ranging international collaboration to define the nature of the last glacial maximum ice cover across the globe was undertaken *c*.40 years ago in a major project known as CLIMAP (Climate: Long Range Investigation, Mapping and Prediction). At the time a key tenet was that the last great ice sheets reached their maximum extent *c*.18,000yr BP. Since then, advances in dating techniques have pushed the real age further back in time. Thus, the interpretation by Richard Fairbanks and Dick Peltier of the Barbados coral reef sea level curve led them to believe that for this area the last glacial maximum may have occurred at least as early as *c*.25,000yr BP. This conclusion was based not only on dating of the Barbados submerged corals, but the results were also internally consistent with Peltier's ICE-5G (VM2) numerical model. These results,

in turn, served to validate Shackleton's marine oxygen isotope chronology, which considered that sea level during the last glacial maximum was around 120–125m lower than present.

Can it be assumed that all of the ice sheets during the last glacial maximum were in glacio-isostatic equilibrium?

This question is nearly impossible to answer! Much depends on how the various low-latitude submerged coral reef records are interpreted (Fig. 8.2). If this curve is accurate there would appear to have been a fall in relative sea level of at least 30m between *c*.30,000 and 25,000yr BP. If *c*.25,000yr BP was the approximate time by which the ice sheets of the northern hemisphere started to reach their maximum extent and thickness, there may have been a *c*.5000yr time interval (between *c*.25,000 and 20,000yr BP) during which time relative sea level did not change appreciably, and the ice sheets maintained

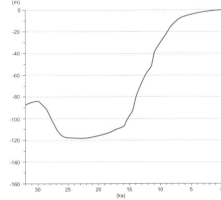

Figure 8.2 Predicted relative sea level history for Barbados using the Peltier ICE-5G (VM2) model. Image credit: Dick Peltier.

their same approximate dimensions. It is simply not known whether or not all of the major ice sheets reached glacio-isostatic equilibrium at this time. As far as the numerical modelling is concerned, glacio-isostatic equilibrium has to be assumed as a starting parameter.

What was the geographical extent of these ice sheets during the last glacial maximum and what were their dimensions – what volume of ice was stored within them?
Many studies have been made over the years to define the nature of ice cover for the last glacial maximum for different areas of the world. Scientists have been able to define the extent and thickness of ice on land and have been doing so for well over a century. For offshore areas, the mapping of ice-marginal moraines on the sea floor is a much more difficult exercise and many offshore areas thought to have been covered by ice during the last glacial maximum have not yet been surveyed in detail. Foremost amongst these are the continental shelf areas surrounding Antarctica.

For many decades scientists have disagreed on the extent of global ice cover for the last glacial maximum (Figs 8.3 and 8.4). Some have advocated a 'minimum' model of glaciation where the ice sheets were restricted to their respective continental shelves and corresponded with an ice volume equivalent sea level lowering of −127m. Others have promoted a 'maximum' model of ice extent, and envisage a much greater ice cover, including numerous marine-based ice sheets that equate to a much greater (up to c.−165m) of sea level lowering. One of the best known

Figure 8.3 Generalised map of ice cover across the northern hemisphere during the last glacial maximum.

areas of the world where there are radically different interpretations of ice cover for the last glacial maximum is Russia. Here, the well-known Grosswald model of 'maximum' ice cover envisages such an extensive former ice cover that much of the meltwater draining from the melting ice is discharged through a cascade of palaeolakes into the Black Sea and ultimately in the eastern Mediterranean.

The analysis of submerged coral reef sequences from areas as diverse as Barbados, the Sunda Shelf, Indonesia, and the Bonaparte Gulf of northern Australia give slightly different results. Whereas the Barbados submerged reefs point to a sea level minimum of c.−120m, the Indonesian and Australian data suggest a greater lowering, possibly as much as −140m. There is no clear consensus on a realistic

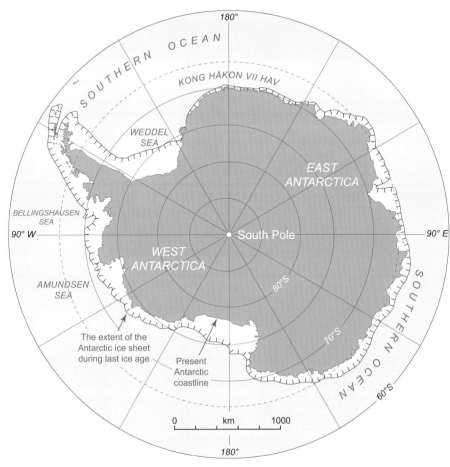

Figure 8.4 One of several possible scenarios of maximum ice cover extent across Antarctica for the last glacial maximum. Image credit: NOAA.

maximum than it is to make comparable estimates for ice sheets. Despite the fact that the edges of numerous ice caps and glaciers terminated in the sea, the fact that most small glaciers and ice caps developed on land makes them easier to measure. Various estimates have been made of their ice age dimensions, and these centre around $c.1.5 \times 10^6 \, \text{km}^3$, equivalent in ice volume terms to a lowering in average sea level of around –4m, a relatively minor amount when compared with the $c.$–120m lowering described above.

Did all the former ice sheets reach their maximum extent at the same time?

Unfortunately, there is no answer to this question. The Barbados coral record, for example, may tell us to what level the sea fell, but it does not tell us anything about the growth and decay history of individual ice sheets. Similarly, the marine oxygen isotope record provides a chronology of ice accumulation and melt over time, but only informs us of average global changes. Logic dictates that in respect of a period of global climate cooling, small ice sheets would build up much faster than large ones. For example, the last British ice sheet is likely to have reached its maximum extent many thousands of years before its North American counterpart. One can envisage that curves of relative sea level change between two such areas are bound to be quite different. During a period of sustained global cooling the small ice sheet would reach its maximum extent while glacio-eustatic sea level was still relatively high. Under such circumstances, the effect of glacio-isostatic depression

ice-equivalent sea level lowering figure, although the balance of opinion leans towards the concept of 'minimum' glaciation and a sea level lowering value closer to –120m. The accuracy or otherwise of these values must always be tempered by the recognition that for specific areas of the world, the local relative sea level lowering is likely to have also been affected by substantial gravitational changes to the sea

surface, as well as by the effects of hydro-isostatic unloading of the ocean floor and the effects of regional and local vertical tectonic land movements.

Can a similar calculation be made for all ice caps and valley glaciers for the last glacial maximum?

In many respects it is easier to calculate the dimensions of glaciers and ice caps that existed during the last glacial could result in exceptionally high relative sea levels around the margins of our theoretically small ice sheet. By contrast, a large ice sheet is likely to have reached its maximum ice extent after glacio-eustatic sea level had fallen considerably, leading to much lower relative sea levels around its margins. This simple description suggests that caution needs to be exercised when interpreting patterns of relative sea level changes in formerly glaciated environments. At the present, there is simply insufficient information to answer this question.

How easy (or difficult) is it to reconstruct palaeo-coastlines for land bridges that existed during the last glacial maximum?

A land bridge represents a connection of land between adjacent continents, between parts of continents, or between islands. Such land connections expanded at times of lower relative sea level and provided routes through which migration and dispersal of plants and animals could take place. It should also be remembered that the evolution of land bridges can also significantly alter regional tidal regimes. During the Late Pleistocene, they enabled faunal interchange and increased species diversity, but equally the disappearance of specific land bridges may have led to faunal extinctions (Fig. 8.5). Since it is estimated that average sea level fell worldwide during the last glacial maximum in the order of –120m, a simple task would be to create a model of the Earth showing the distribution of land masses that arises when sea level is lowered by this amount. Caution should be

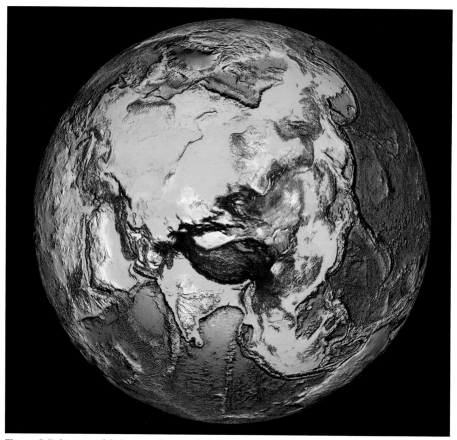

Figure 8.5 Image of Asian continent giving general indication of potential areas of land bridges that may have existed during the last glacial maximum. Image credit: National Geophysical Data Center at NOAA.

exercised, however, since distinction should be made between different types of land bridges that existed during the last glacial maximum.

Land bridges located distant from the northern hemisphere ice sheets (farfield areas)

The largest and best known of these is the Sunda archipelago (Fig. 8.5). At first sight it is a simple exercise to lower sea level by –120m and construct a map showing the newly exposed land

areas. However, vertical tectonic movements of the Earth's crust also become an important factor in determining the positions of last glacial maximum coastlines, with former coastline positions varying according to how much uplift or subsidence there has been over the last c.20,000yr. An additional factor to consider is how much regional sea level was altered by changes in the geoidal sea surface during the last glacial maximum. The recognition that such processes may have taken place would suggest that caution needs to be exercised before reconstructing palaeo-coastlines and identifying the key land bridge routes. At the very least, shoreline positions created as a result of geophysical modelling would need to be compared with empirical field data before any land bridge in this area could be reconstructed with a high degree of confidence.

Land bridges in the northern hemisphere located beyond the limits of large ice sheets

For the northern hemisphere the most important land bridge was Beringia, which connected North America and Eurasia (Fig. 8.6). At this time Beringia was a polar desert flanked to the east by small ice caps over Kamchatka and to the south by an ice cap across southern Alaska and the Aleutian islands. This ice cap was continued to the west and south by the Cordilleran ice sheet which, in turn, merged with the Laurentide ice sheet. The eastern part of Beringia across central and western Alaska remained ice free apart from an ice cap located over the Brooks Range, central Alaska.

The most common method of reconstructing the palaeogeography of Beringia for the last glacial maximum has been to define the former coastal margin at the –120m contour. However, the precise position may be more difficult to determine, since there may have been significant regional variations in patterns of glacio-isostatic adjustment. These include resolving: a) which areas of landscape beyond the respective ice sheets and ice caps were affected by peripheral forebulge depression; b) where did any zones of forebulge uplift

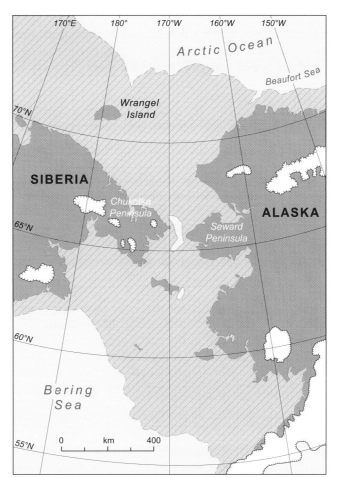

Figure 8.6 Reconstruction of Beringia for last glacial maximum. Image credit: United States Geological Survey.

extend to; c) what gravitational attraction of ocean water to ice masses took place in this area; and d) were any areas affected by vertical co-seismic and inter-seismic shoreline displacement?

Small land bridges covering a limited geographical area

Thousands of such areas occur throughout the world, in many cases represented by groups of small islands no more than a few kilometres apart.

9 Ice sheets and meltwater pulses

Introduction

The melting histories of the various ice sheets on Earth following the last glacial maximum and the associated rise in global sea level are hotly debated topics amongst scientists. But why? Surely, since these changes took place thousands of years ago, they must be of little consequence to our understanding of present patterns of change? In fact, there are several important reasons.

One reason is that the end of this glacial period marked the start of a process of glacio-isostatic readjustment as the lithosphere was released from beneath the masses of overlying ice, in particular from the largest ice sheets that had covered extensive areas of North America and northern Europe. Remarkably, this process of glacio-isostatic adjustment is still continuing today, over 20,000 years later. The most obvious manifestation of this is in areas close to the centres of these former ice sheets, where the ice thicknesses were greatest. For example, tide gauge records in northern Scandinavia along the shores of the Baltic Sea demonstrate very clearly that relative sea level continues to fall in these areas ($c.7$myr^{-1}) as the rate of glacio-isostatic rebound continues to outpace the glacio-eustatic rise caused by the increased global ocean mass and volume. In areas located beyond the former ice sheets, the continuing collapse of crustal forebulge areas is an important factor contributing to relative sea level rise. These two patterns of change tell us, above all else, that we need to be able to understand the nature of these glacio-isostatic adjustments, since these have to be distinguished from the effects on sea level of recent climate change.

The melting histories of the last great ice sheets also enable the reconstruction of past rates of glacio-eustatic sea level rise. This may be significant from the point of view of understanding how the Earth's oceans might respond in the future to different scenarios of ice melt. At present there is concern about the impact on sea level rise of ice melt in Greenland and across West Antarctica (see chapter 11). Is it possible that parts of these ice sheets could melt in the future and, if so, what sorts of future sea level might we expect? A key step in this process of understanding is to learn more about how fast glacio-eustatic sea level rose in the past in response to the melting of the last great ice sheets. Can they be considered as analogues for potential future changes in sea level?

Defining timescales

A cursory inspection of some of the geological literature highlights a range of nomenclature used to define the time interval between the last glacial maximum and the start of the Holocene epoch. The period of time that started with the melting of the ice sheets across the northern hemisphere and ended with the beginning of Holocene warmth is often referred to as the last glacial–interglacial transition. It is a complex period of time to interpret, but broadly it can be subdivided into three parts. The first and oldest is the period of time during which the ice sheets were melting, sea levels were rising – sometimes very rapidly – yet there were no clear indications that air or ocean temperatures were rising markedly (table 9.1). A second period of change appears to have commenced $c.$14,000yr BP with clear indications of a rise in sea surface temperatures. This period of climatic amelioration is referred to as the Lateglacial Interstadial, and appears to have lasted until $c.$12,900yr BP, when there was renewed cooling. This latter cold period is generally known as the Younger Dryas and is thought to have ended abruptly $c.$11,700yr BP when it was replaced by the start of the Holocene interglacial.

In the context of sea level change research, the subdivision of time between 'Lateglacial' and 'Holocene' does not matter all that much, since many of the key processes of glacio-isostatic adjustment are relatively insensitive to patterns of global climate change. There is a need for caution, however, since the coral reef records point to the Younger Dryas period of cold climate as having been both preceded and

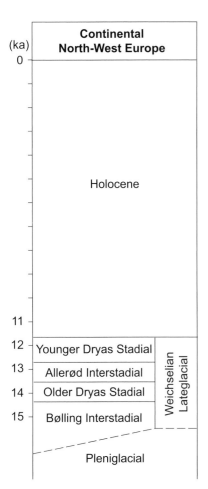

Table 9.1 Generalised chronology for Lateglacial across NW Europe. Image credit: adapted from J.J. Lowe and M.J.C. Walker (2014).

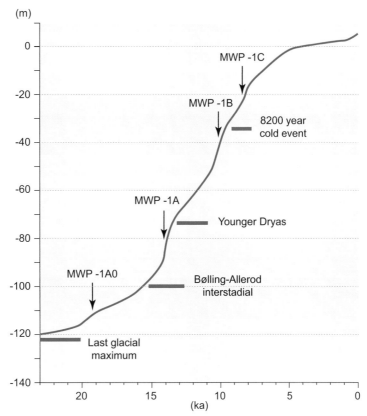

Figure 9.1 Timing of global meltwater pulses since the last glacial maximum based on analysis of submerged coral data. Image credit: NASA, Vivien Gornitz.

followed by periods of very rapid global sea level rise (known as meltwater pulses). The occurrence of these, and earlier, meltwater pulse events were first described from the Barbados sea level record in 1995 by Paul Blanchon and John Shaw, who described them as catastrophic rise events (Fig. 9.1). Blanchon and Shaw also highlighted other time intervals (e.g. the Younger Dryas)

when the rate of sea level rise appears to have slowed down. But the sequence of meltwater pulses appears to have started many thousands of years earlier. Moreover, some of reconstructed rates of sea level rise for these time intervals appear to have been very high indeed.

Meltwater pulses

The submerged coral reef stratigraphy from Barbados indicates that between c.19,000 and 7000 years ago glacio-eustatic sea level rose by c.110m (Fig. 9.1). This change is broadly equivalent to an average rate of rise for this

period in the order of 1m per century, or expressed differently as 10mmyr^{-1}. However, at specific time intervals during this period, glacio-eustatic sea level rose at much faster rates, sometimes in excess of 40mmyr^{-1}. The first of these meltwater pulses, (referred to as a catastrophic rise event, CRE1) by Blanchon and Shaw, occurred c.19,000yr BP and probably represents an initial acceleration in ice-sheet melting following the last glacial maximum. This event is known more commonly in the scientific literature as meltwater pulse 1A0. Although we cannot be sure of the sources of the melting ice, it is thought likely that a significant proportion of the meltwater was derived from the thinning of ice sheet surfaces. A popular estimate is that during this meltwater pulse, sea level rose between *c.*16 and 25m in *c.*400–500 years.

That the process of glacio-isostatic rebound can begin without any significant change in the position of the ice sheet margin, and mostly due to ice-sheet thinning, was elegantly demonstrated by John Andrews in the early 1970s. Andrews demonstrated that the thinning of the surface of an ice sheet during melting is associated with an initial restrained glacio-isostatic rebound of the lithosphere. He showed that as much as half of the total glacio-isostatic rebound of the lithosphere in a formerly glaciated area could occur while ice still covered the landscape.

The second meltwater pulse (MWP 1A) is perhaps more spectacular than the first, and may also have been intimately related to ice-sheet thinning (Fig. 9.1). This rise occurred sometime between *c.*14,600 and 13,500yr BP and was first recognised from the Barbados coral record. This period of rapid rise was in the order of 20–30m and, as with meltwater pulse 1A0, equated to an average rate of rise in the order of *c.*20m over *c.*500 years, approximately equal to a rate of average sea level rise of between 40 and 60mmyr^{-1}. Scientists are unclear from which ice sheet/s this meltwater originated. Some have argued that it was mostly derived from widespread melting across Antarctica, while others have proposed that the source of the meltwater was predominantly from the Laurentide ice sheet.

In attempting to recognise the former occurrence of periods of accelerated sea level rise such as MWP 1A, it is important to remember that sometimes the field evidence for the occurrence of a former meltwater pulse can be misleading. For example, in formerly glaciated areas, a period of renewed ice sheet buildup following a climatic deterioration can induce a period of glacio-isostatic depression that interrupts a period of glacio-isostatic rebound. In such circumstances, the coastal stratigraphic data from a field site located close to a former ice sheet can give the appearance of a meltwater pulse having taken place when none in fact took place. Fortuitously, all of the coral reef chronologies used to reconstruct past sea level are for locations in the farfield, and unaffected by vertical glacio-isostatic movements.

A later meltwater pulse (MWP 1B) is thought to have taken place at the start of the Holocene following the end of the Younger Dryas period of cold climate (Fig. 9.1). This period of accelerated sea level rise, sometimes known as catastrophic rise event CRE2, was first recognised from the Barbados coral reef record analysed by Richard Fairbanks. Opinion is more divided on the existence of this meltwater pulse. Fairbanks and co-researchers maintained that the rise took place around 11,300yr BP and that it represented a similar order of rise (*c.*20m in *c.*500 years) to earlier meltwater pulses. They argued that an acceleration of glacio-eustatic sea level rise at this time was probably a response to the global climate warming that followed the termination of the Younger Dryas period of cold climate. However, research on

submerged coral reef sequences off Tahiti undertaken by Eduard Bard and his team did not show the same patterns of change. Bard maintained that the Tahiti sea level record for this time period, despite pointing to a sustained rise, did not indicate that a sudden jump in sea level had taken place sufficient to be classified as a meltwater pulse. There are also other interpretations. For example, J Paul Liu and John Milliman undertook a re-investigation of the Barbados and Tahiti sea level data and argued that there had indeed been a very rapid rise in glacio-eustatic sea level between 11,500 and 11,200yr BP in the order of *c.*12m equivalent to an average rate of rise of *c.*40mmyr^{-1}. Other studies have contested this assertion and have downgraded the estimated amount of rise by almost half.

As with the earlier meltwater pulses, it is almost impossible to know where this glacial meltwater came from. Much debate concerns the nature of the period of cold climate, the Younger Dryas Stadial, that preceded MWP 1B. Evidence for this period of exceptionally cold climate spanning *c.*1300 years appears in numerous ice cores and ocean sediment cores. It is well known that this period of cold climate was particularly severe across NW Europe, while advance of glaciers may have taken place elsewhere around the world. It is thought that at this time the ice sheet across Scandinavia, having been subject to retreat and thinning over many millennia, was subject to renewed ice accumulation – only to suffer later accelerated melting with the climatic warming that marked the start of the Holocene interglacial. Whether the ice sheets in North America, Greenland

and Antarctica were subject to renewed ice buildup during the Younger Dryas is an open question. Whatever buildup of ice did take place at this time, it was not sufficient to halt the sustained rise in sea level that took place between meltwater pulses 1A and 1B.

The final meltwater pulse (MWP 1C) described by scientists is perhaps the most complicated of all. The coral reef record points to a 'jump' in sea level between c.8200 and 7600yr BP (Figs 9.1 and 9.2). Complications arise because this time interval coincided with the final disintegration of the Laurentide ice sheet. By contrast, the final melting of the Eurasian ice sheet was nearly complete by c.8000yr BP, with just fragments of a small ice cap remaining across northern areas of Norway, Sweden and Finland. For the most part the final melting of ice across Scandinavia and European Russia took place between c.11,000 and 9000yr BP. Given the enormous size of the Laurentide ice sheet, the melting of this ice sheet took place over a longer time interval. For example, by c.8400yr BP, a large ice sheet still covered much of central and eastern Canada, yet by c.7600yr BP most of this had melted (Fig. 9.2). Although little is known about mass balance changes for the Antarctic and Greenland ice sheets around this time period, it is reasonable to argue that the final stages in the melting and disappearance of the Laurentide ice sheet contributed most to the timing and size of meltwater pulse 1C.

Emptying of glacial lake Agassiz-Ojibway and sea level change

One additional period of rapid sea level rise should be mentioned. During the thinning and retreat of the southern margin of the Laurentide ice sheet, an enormous body of water became dammed against the southern edge of the ice sheet (Fig. 9.2). At its maximum size, the lake, known as glacial lake Agassiz-Ojibway, extended over 3000km along the southern edge of the ice sheet and had a volume in the order of c.150,000km³. The main overflow route was in the west, where the water drained southward to the west of the incipient Great Lakes. Several years ago, Jim Teller demonstrated that as ice-sheet melting continued, the lake was suddenly emptied catastrophically, possibly over several days.

The process by which the lake was drained has been debated for many years, with the generally agreed explanation that the lake waters drained northwards underneath the ice sheet

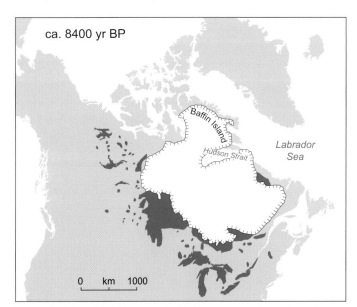
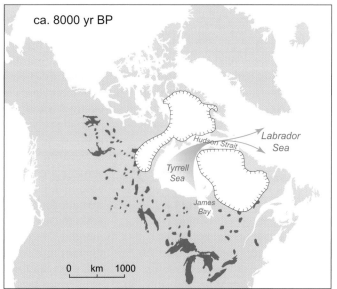

Figure 9.2 Glacial lake Agassiz-Ojibway drained northward through a network of subglacial tunnels underneath areas of the Laurentide ice sheet. The lake waters emptied into the North Atlantic through the Hudson Strait. Following the disintegration of the ice sheet, marine waters flooded westwards through the Hudson Strait into Hudson Bay, forming the Tyrrell Sea.

through a network of subglacial tunnels. The water emerged in the northern part of what is now Hudson Bay, and thereafter flowed eastwards catastrophically into the North Atlantic through the Hudson Strait (Fig. 9.2). Many aspects of the reconstructions of what happened have been worked out by Arthur Dyke and co-workers at the Canadian Geological Survey of Canada. The scale of the flooding is also indicated by large iceberg scour marks that have been found on the floor of the Hudson Strait. Once most of the water had escaped from beneath the ice as a series of jokulhlaupe (the Icelandic term to describe such events), the remaining masses of ice that were located across Labrador and the Hudson Bay area are thought to have collapsed in the ensuing centuries, causing the final disintegration of the Laurentide ice sheet. Several hundred years later, nearly all of the ice had disappeared from the North American landscape.

So what is the relevance of this massive rapid meltwater discharge through the Hudson Strait into the North Atlantic? The answer is that the sudden emptying of glacial lake Agassiz-Ojibway implies that a huge volume of freshwater was discharged into the North Atlantic within a very short period of time. Our knowledge of modern jokulhlaup events suggests that the tunnels beneath the ice themselves became enlarged quickly as a result of the transfer of sensible heat from water to the surrounding ice walls. Many consider that the entire lake could have been emptied within days or weeks. If the estimated volume of lake water of 150,000km^3 is anywhere close to being accurate, it implies that glacio-eustatic sea level could have risen by up to c.0.5m within a matter of days or weeks – indeed, some have argued for a much greater rise.

This sort of estimate of near-instantaneous sea level rise is astonishing. It contrasts with all of the meltwater pulse events, where much greater amounts of glacio-eustatic sea level rise took place over centuries. The environmental changes that took place in association with the Agassiz-Ojibway flood effectively marked the last of a chain of events that led to the disintegration of the Laurentide ice sheet. One may safely assume that these changes, together with the earlier meltwater pulses, were accompanied by changes in the distribution of ocean mass across the surface of the planet. Equally, the disappearance of the Laurentide and Fennoscandian ice sheets may have led to profound changes in the flow of material within the Earth's mantle. With these changes came other effects. First, the sudden addition of such a huge volume of freshwater to the world's oceans created pressures (hydro-isostatic) on the underlying oceanic lithosphere. Second, the glacio-isostatic rebound across ice-covered landscapes was accompanied by the collapse of glacio-isostatic forebulge regions across huge areas outside of the former ice sheet margins. Third, the disappearance of the northern hemisphere ice sheets led to gravitational changes and a lowering of sea level across ocean areas surrounding the former ice sheets. As seawater was transferred away from the former ice sheets in this manner, the displaced ocean water was transferred towards farfield ocean areas. When all of these processes are considered together, it becomes clear that the patterns of relative sea level change that accompanied the melting of the last great ice sheets were highly complex – indeed, as a result of a combination of factors, relative sea level may have fallen across some areas where one might have expected a rise to have taken place – and vice versa.

Introduction

In contrast to the meltwater pulses that accompanied the melting of the last great ice sheets, the changes in relative sea level that took place after the ice sheets in the northern hemisphere had melted were on a much smaller scale. It is estimated that nearly all remnants of the Laurentide ice sheet had disappeared from Canada by c.7000yr BP. It can also be stated with confidence that the Fennoscandian ice sheet had already melted several centuries earlier. Of course both the West and East Antarctic sheets plus the Greenland ice sheet remained intact and resembled something similar to their present dimensions. Inspection of the curves of Holocene sea level change from around the world all tell the same story – that regional rates of sea level rise slowed down drastically after c.7000yr BP.

Was it the case that after the enormous ice sheets in North America, Scandinavia and Russia had disappeared there were no large ice sheets left to melt? Of course, the exceptions to this explanation are the ice sheets in Greenland and Antarctica, but from what we can tell they did not experience any major shrinkage in size during the middle and late Holocene. For the most part, the patterns of relative sea level changes that took place over the last c.7000 years were mostly responsive to fluctuations in the sizes of the world's c.150,000 valley glaciers and small ice caps rather than to any major changes in the surface mass balances of the Greenland, West Antarctic and East Antarctic ice sheets. Considered simply, one might expect that perturbations in global climate during the middle and late Holocene may have caused alpine glaciers and small ice caps to expand and contract in response to episodes of global cooling and warming. However, glaciers in different parts of the world responded differently to past change in Holocene climate, with periods of climate cooling in one part of the world accompanied by warming in another. Furthermore, there is no particular reason to think that patterns of Holocene climate change that took place across the northern hemisphere would have been replicated across the southern hemisphere. It follows, therefore, that one might expect that patterns of relative sea level change during the middle and late Holocene may have been relatively minor.

The abandonment of the concept of global sea level curves

During the latter part of the twentieth century, Rhodes Fairbridge, in a series of classic papers, tried to reconstruct patterns of sea level change from around the world that he considered were 'global' in nature. He made used of sea level change data from Australia, together with information from the northern hemisphere, to make the case for an oscillating curve of Holocene sea level change. The 'Fairbridge eustatic sea level curve' also depicted sea level having reached a level higher than present throughout the majority of the Late Holocene. This sea level curve contrasted markedly with that produced by Francis Shepard in 1961, which depicted a smooth and decelerating rise in sea level throughout the Holocene with none of the 'oscillations' later envisaged by Fairbridge. Discussion of the merits of these and other descriptions of the nature of Holocene sea level change stimulated often fierce debate from the 1960s onwards for several decades, with the central issue of controversy being whether or not the so-called 'global' eustatic rise in sea level was smooth or oscillatory (Fig. 10.1).

During those years a Swedish geoscientist, Nils-Axel Morner, repeatedly expressed the argument that geoidal sea surface processes made it obvious that there was no such thing as a global eustatic sea level. His scientific arguments represented a paradigm shift in how we understand and interpret research on past changes in relative sea level. The reasoning that he used to demonstrate that the concept of a global eustatic sea level curve was an illusion was supported by results from numerical modelling exercises that considered the various long-term

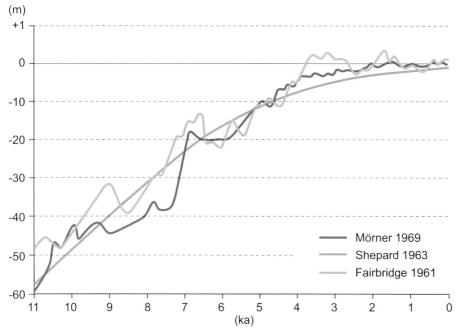

Figure 10.1 Schematic representation of three competing graphs of Holocene relative sea level change, each promoted by different groups of research scientists during the late 20th century.

responses of the Earth's sea surface to the melting histories of the ice sheets. One of the first such studies was that of RI Walcott, who developed a simplified model of the Earth to estimate its elastic deformation due to ice sheet melting, coupled with what was known at the time about the associated glacio-eustatic increase in ocean volume, which Walcott referred to as 'a time-dependent set of sea level altitude measurements averaged for all of the world's oceans'.

One of Walcott's main conclusions was that one could quite simply account for the different reconstructions of Holocene sea level change for different locations around the world through a consideration of the effects on flow within the mantle of changes in the distribution of water and ice loads across the Earth's surface. Morner's research paid particular attention to the way that long-term gravitational changes led to changes in the geoidal sea surface. For example, he highlighted the counter-intuitive process that the melting of a large ice sheet led to a reduction in the gravitational attraction of ocean water to ice sheets, and that as an ice sheet continued to melt, sea level in ocean areas surrounding the ice sheet would fall rather than rise.

During recent decades these pioneering ideas have been developed into highly sophisticated numerical models. Foremost amongst those who have explored these issues in considerable detail have been Dick Peltier, Kurt Lambeck, Glenn Milne, Jerry Mitrovica and John Chappell. Their studies have shown unequivocally that the effects of glacio-isostatic adjustment on patterns of sea level change extend into farfield ocean areas. In other words, the reconstructions of past sea level from individual locations using geological evidence frequently depart, often considerably, from what might be expected if the global oceans shifted uniformly up and down in response to ice melt (i.e. as a result of a glacio-eustatic change). A classic illustration of such changes having occurred is for Australia, where patterns of regional sea level change are strongly influenced by the response of continental shelf areas to changing water loads during the Holocene. These changes are referred to as hydro-isostatic adjustments, which reflect the ways in which continental shelf areas have deformed in response to changes in water loads. John Chappell showed how continental shelf margins are downwarped as a response to lower relative sea levels across areas of the outer shelf. At the same time, coastal areas experience upward flexure as part of what Chappell described as a 'hinge' type of motion that results in relative sea levels across inner shelf areas being higher than they otherwise would be. During the latter part of the Holocene this has led to a long-term lowering of relative sea level across continental shelf areas of Australasia, with the dimensions of this lowering having varied according to the width of individual continental shelf areas.

The above example from Australia serves to illustrate how sea level change research has experienced a paradigm shift during recent decades. The long-established concept of a single eustatic sea level curve for the whole world has been abandoned completely. In its place has come the realisation that there are a multiplicity of geological processes that influence patterns and rates of relative sea level changes for any particular location. Foremost amongst these are the effects of glacio-isostatic adjustment that includes not only glacio-isostatic deformation across areas covered by former ice sheets, but also the surrounding zones affected by crustal forebulge and collapse, as well as farfield effects. Add to this the changes in the gravitational attraction between oceans and adjacent ice sheets and related water displacement both towards and out of the farfield, as well as processes of hydro-isostatic deformation, and one need not be surprised that the idea of a single global eustatic sea level curve has disappeared.

Regional variations in patterns of Late Pleistocene and Holocene relative sea level change

In this section several examples are given of different regional patterns of relative sea level that took place during the Late Pleistocene and Holocene. The patterns described are accompanied by explanations of the key geological and oceanographic processes that make each type of relative sea level curve distinctive. In this section attention will focus on the key processes responsible for particular patterns of change in different regions, rather than considering the finer details of any particular published sea level curve. To this end, summaries are given here of past patterns of sea level change characteristic of:

◆ emerged coastal landscapes located close to the central areas of former ice sheets;
◆ coastal landscapes located within areas covered by ice during the last glacial maximum but closer to former ice sheet margins;
◆ coastal landscapes located close to the former edges of ice sheets yet within the ice sheet margin;
◆ coastal landscapes located within crustal forebulge zones; and
◆ coastal landscapes in farfield regions.

Emerged coastal landscapes located close to the central areas of former ice sheets
North America
Hudson Bay
In North America, the final disintegration of the Laurentide ice sheet was accompanied by the incursion of seawater from the North Atlantic into Hudson Bay to produce what is described as the Tyrrell Sea (Fig. 9.2). As one might expect, the sea did not flood across the whole of Hudson Bay simultaneously, with some areas remaining ice-covered longer than others. As a consequence of glacio-isostatic rebound, a succession of shorelines was produced around Hudson Bay and parts of the Canadian Arctic archipelago (Fig. 10.2). Some of the raised shoreline features are located as far as several hundred kilometres inland and up to c.300m above sea level. Many of these shorelines are preserved in today's landscape as uplifted beach ridges. In some areas spectacular flights of beach ridges cover individual hillslopes. Dating

Figure 10.2 Emerged Holocene beach ridge staircase, Bathurst Inlet, Nunavut, northern Canada. Image credit: Mike Beauregard, Creative Commons.

of the oldest marine shells from the highest (and thus oldest) beach ridges appear to indicate that the Tyrrell Sea flooded across Hudson Bay soon after *c*.8300yr BP. It is known that the pattern of regional ice sheet decay was asymmetric. Rather than the simple disintegration of an ice dome over Hudson Bay, the distribution of elevations of the oldest emerged shoreline fragments tend to suggest the former existence of separate domes of uplift both to the W and NE of Hudson Bay.

The reconstructions of relative sea level change for this area of North America all take the form of decay curves that mirror a trend of an exponential decrease in glacio-isostatic uplift over time. An important observation, however, is that the relative sea level curves only cover a relatively short period of time (*c*.8300yrs) since prior to this time the area was covered by ice, although glacio-isostatic rebound may have been initiated as much as *c*.10,000 years earlier. Inspection of Figure 10.2 might leave the impression that the pattern of falling relative sea level is simply due to decelerating glacio-isostatic rebound. However, this view is incorrect since the influx of ocean water into Hudson Bay at *c*.8300yr BP took place when the ice volume-equivalent sea level was *c*.−30m lower than at present. This means that a raised shoreline in the Richmond Gulf with an altitude of +180m and an age of *c*.8300yr BP has experienced +210m of uplift over this time interval. Matters are further complicated by a relative lowering of the geoidal sea surface caused by the disintegration and final disappearance of the Laurentide ice sheet over Hudson Bay.

Vancouver Island

During the last glacial maximum the Laurentide ice sheet merged to the west with the Cordilleran ice sheet that extended from southern Alaska to northern Washington and Montana. Together the two ice sheets covered *c*.4000km from east to west, and reached several kilometres in thickness with an uneven topography characterised by multiple domes and saddles. Under such circumstances it is not surprising that the pattern of deglaciation was complex and intricate. This was certainly the case in respect of the Cordilleran ice sheet. At an early stage during deglaciation the ice thinned and retreated sufficiently to create a land corridor between it and the western margin of the Laurentide ice sheet. Along the Pacific coast the pattern of relative sea level changes that accompanied ice thinning and melting was not only affected by glacio-isostatic rebound but also by vertical tectonic movements. For example, research by Tom James, Ian Hutchinson, John Clague and colleagues in eastern Vancouver Island has demonstrated that the earliest incursions by the sea occurred soon after *c*.14,000yr BP when it may have reached *c*.+150m. Thereafter, relative sea level fell steadily until *c*.11,500yr BP to a level of *c*.−15m (Fig. 10.3).

This remarkable fall in relative sea level is for the most part attributable

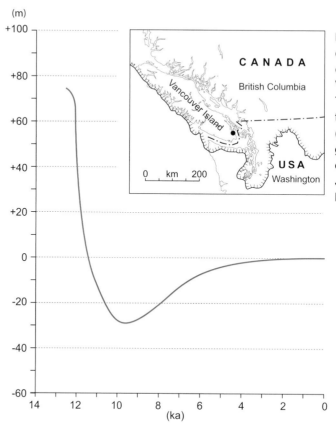

Figure 10.3 Trend of relative sea level change, southern Vancouver Island. The solid line depicts the extent of the last (Late Wisconsinan) glaciation. Image credit: Thomas James and Ian Hutchinson.

to glacio-isostatic rebound. However, one must also remember that the coral reef records tell us that ice volume sea level equivalent at 14,000 years ago was in the order of *c*.–100m, thus implying that over the succeeding *c*.3000 years the total glacio-isostatic uplift across eastern Vancouver was in the order of 265m (150 + 15 + 100m). To this we need to make additional adjustments in respect of geoidal lowering of the adjacent Pacific Ocean surface as ice retreat progressed, as well as vertical tectonic land movements across a well-known area of active plate collision. James and his team also showed, from isolation basin studies, that at the start of the Holocene, relative sea level rose to *c*.+1m between *c*.9000 and *c*.8500yr BP and thereafter remained close to its modern-day level. Other studies of relative sea level change in different parts of Vancouver Island have revealed slight differences in the deglaciation history, as well as regional differences in the highest levels reached by the sea as the ice sheet melted. All of the studies, however, show the same marked lowering of relative sea level during the Lateglacial and Younger Dryas to reach levels below present sea level (sometimes as low as *c*.–30m) just prior to the start of the Holocene.

The patterns of relative sea level observed for Vancouver Island (early lowering, lowstand, later rise, then gradual decline until present) is one that also occurs in other formerly glaciated environments of the world (e.g. Iceland, Scotland). Broadly speaking, distinctive sets of processes that help define the observed patterns of relative sea level change can be grouped into four time intervals:

◆ First, the early lowering of relative sea level is principally due to rapid regional glacio-isostatic rebound that outpaces the rate of rise of glacio-eustatic sea level caused by the melting of ice worldwide (the ice-equivalent sea level rise described earlier). As stated earlier, glacio-isostatic rebound along ice-free coastal areas is accompanied by restrained glacio-isostatic rebound across areas still covered by ice. Although there is a tendency to make a distinction between glacio-isostatic rebound across ice-free areas and ice-covered areas, in practice there is a continuum of differential isostatic rebound across the entire region.

◆ Second, the lowstand of relative sea level that occurs along the Pacific west coast during the latter part of the Younger Dryas Stadial represents a period of time when the glacio-eustatic rise in sea level (increase in ocean volume due to ice melt worldwide) takes place at the same approximate rate as the (decreasing) rate of glacio-isostatic uplift. This is the main reason why no apparent rise in relative sea level takes place at this time.

◆ Third, the subsequent rise in relative sea level at the start of the Holocene is due to the rate of ocean mass increase (glacio-eustatic sea level) overtaking the rate of (decreasing) glacio-isostatic rebound. Remember that the beginning of the Holocene marks a period of time when the last great ice sheets were melting rapidly. Although the last remnants of the Laurentide ice sheet across Labrador had not melted completely until *c*.7000yr BP, most of the Cordilleran ice had disappeared by the start of the Holocene.

◆ Fourth, the slight lowering of relative sea level across the Pacific coast of Canada between *c*.8000 and *c*.7500yr BP until present can be explained in a number of ways. First, if it is assumed, as is the case elsewhere, that glacio-isostatic uplift is still incomplete, one might explain the progressive lowering as the result of continued glacio-isostatic rebound. Another explanation might be that the floor of the Pacific Ocean may have deepened slightly during the mid-Holocene. It might also be the case that vertical tectonic land movements have interfered with the reconstructed sea level histories. Lastly, it may also be the case that long-term gravitational changes resulted in deformations of the geoidal sea surface and may have played an important part in determining local relative sea level histories.

Scandinavia

Ever since the classic Swedish research of Gerard de Geer in the late nineteenth century, patterns of relative sea level change for Scandinavia have been investigated in considerable detail, perhaps more than any other area of the world. The distribution of raised shoreline features of Late Pleistocene and Holocene age across Scandinavia display a broad pattern of differential isostatic rebound, with the reconstructed uplifted isobases positioned around a central dome of maximum uplift across northern Sweden. Reconstructions of former patterns of relative sea level are particularly complex for landscapes surrounding the Baltic Sea basin, owing to the interaction of glacio-eustatic and glacio-isostatic processes that resulted in the development of several freshwater lakes and marine incursions throughout the Lateglacial and Holocene. Svante Bjorck demonstrated that, as the ice sheet across Scandinavia thinned and retreated, a large lake, the Baltic Ice Lake, developed along the southern and SE margin of the retreating ice sheet and overflowed into the North Sea near Mt Billingen (Fig. 10.4). Near the end of the Younger Dryas period of cold climate, ice retreat caused the Baltic Ice Lake level to fall by *c.*25m. During this time *c.*7–8000km^3 of freshwater was drained out of the lake into the North Sea. Several hundred years later, when the rate of glacio-eustatic sea level rise outpaced the rate of land uplift in the southern Baltic, the area experienced a marine incursion that resulted in the development of the Yoldia Sea, (named after the cold-water marine shell *Portlandica (Yoldia) arctica*) (Fig. 10.4). Thereafter, by *c.*10,200yr BP, continued glacio-isostatic uplift across southern Sweden led to the closure of the marine connection and the development of the freshwater Ancylus Lake. During the early Holocene, when the rate of glacio-eustatic sea level rise exceeded the rate of crustal rebound in southern Scandinavia, brackish water started to invade the lake and between *c.*9–8500yr BP marine waters started to spread across the Baltic. This relative marine transgression, known as the Litorina transgression, lasted *c.*3000 years, converting the Baltic to fully marine conditions (Fig. 10.4).

The series of 'switches' from (Baltic Ice Lake) freshwater to (Yoldia) marine, to (Ancylus) freshwater to (Litorina) marine represent a classic illustration of the interplay between glacio-eustatic

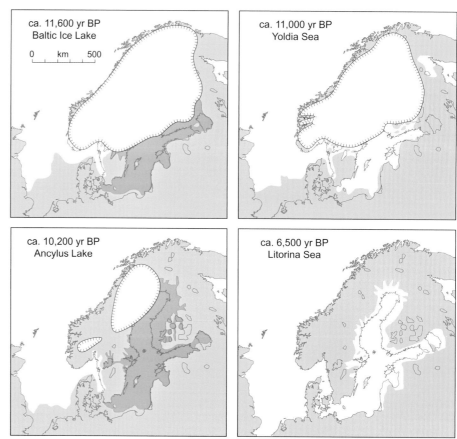

Figure 10.4 Evolution of Baltic Ice Lake, Yoldia Sea, Ancylus Lake and Litorina Sea caused by interaction of glacio-eustatic and glacio-isostatic changes.

and glacio-isostatic changes (Fig. 10.4). They also show how regional patterns of glacio-isostatic rebound across Scandinavia were complicated by the repeated loading and unloading of water masses and associated gravitational changes. The reconstructed patterns of relative sea level change across Scandinavia were summarised by Jan Mangerud and his colleague Jon-Inge Svendsen, who constructed a shoreline height–distance diagram to illustrate these changes covering the last *c.*13,000 years (Fig. 10.5) They produced a

curve of relative shoreline displacement for western Norway along a line perpendicular to the uplift isobases, based on detailed geomorphological mapping as well as on the results of numerous studies of isolation basins (see chapter 4). They illustrated how all emerged shorelines decline in altitude westwards perpendicular to the uplift isobases, with younger shorelines exhibiting successively lower regional tilts, this indicative of decreasing rates of glacio-isostatic rebound with time. Their studies showed that the majority

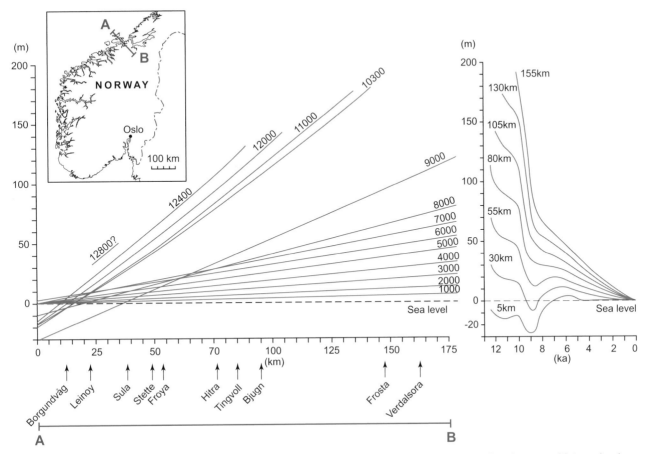

Figure 10.5 Suite of theoretical relative sea level curves for western Norway reconstructed at uniformly spaced intervals along shoreline height–distance diagram. Image credit: Jan Mangerud and Jon-Inge Svendsen.

of the Lateglacial was dominated by crustal emergence, which slowed down to a near stillstand during the Younger Dryas period of cold climate. Emergence resumed during the early Holocene until punctuated by the Tapes (equivalent to the Litorina transgression in Sweden) relative marine transgression, which culminated during the mid-Holocene at *c.*6000yr BP and which was followed by a relative marine regression (Figs 10.4 and 10.5).

The shoreline height–distance diagram for western Norway is important for another reason. Owing to the fact that the ages of several of the tilted shorelines are known, it was possible to derive a series of relative sea level curves from this data. This was done by plotting the respective ages and altitudes of each shoreline at equal distances along the x-axis of the height–distance diagram to produce a series of theoretical relative sea level curves (Fig. 10.5). The sea level curves provide an excellent example of how

the shape of a relative sea level curve varies according to distance from the centre of glacio-isostatic uplift. It should be noted that the numerical models of glacio-isostatic adjustment can also produce this information as an output, provided that the relevant sea level index points are included as part of the modelling exercise.

Scotland

The methodology used to reconstruct former patterns of relative sea level change for Scotland differs in some major respects from that used in Scandinavia. The most important difference is that isolation basin methodology, introduced by Ian Shennan and his colleagues at Durham University, has only been used during the last 20–30 years to reconstruct past patterns of relative sea level changes. By contrast, much of the early research on former sea level changes has focused on stratigraphic and morphological evidence for past changes in relative sea level. In particular, the pioneering research led by Brian Sissons focused on the Lateglacial and Holocene stratigraphy of the Forth Valley in eastern Scotland. His shoreline height–distance diagram shows a series of glacio-isostatically tilted shorelines that are continuous over many tens of kilometres. Although uplifted features, many of the reconstructed shorelines occur beneath the ground surface and are broadly tilted from west to east with the oldest (Lateglacial) shorelines exhibiting the greatest amounts of differential tilting. These palaeoshorelines are distinctive from their counterparts in Norway in two main respects. Firstly, shorelines of similar ages occur at much lower altitudes in Scotland than in Norway, this reflecting much smaller amounts of glacio-isostatic rebound and a thinner ice sheet in Scotland than in Scandinavia. For example, the highest Lateglacial raised shorelines in eastern Scotland occur at c.40m above sea level while features of a similar age in Norway occur up to c.200m above present sea level. The second difference is that, in contrast to Scandinavia, the ages of the tilted shorelines in eastern Scotland are not well dated. If they were, then it would

be possible to produce a series of curves of relative sea level change perpendicular to the shoreline uplift isobases for eastern Scotland in the same way as has been accomplished in Norway.

For the majority of mainland Scotland the reconstructed curves of relative sea level change exhibit similar patterns of change to those in Norway except on a much smaller scale. Thus, there is an initial fall when the rate of rebound outpaced the rate of glacio-eustatic rise which, in turn, may have been countered by gravitational lowering of the sea surface due to ice melt. This was followed by a lowstand when the rebound rate and the eustatic rise rate were approximately balanced. This was succeeded during the early Holocene by a relative rise, when the glacio-eustatic rise exceeded the rate of land rebound. Finally, the last phase occurred during the latter part of the Holocene when decelerating land uplift continued during a period of time when glacio-eustatic sea level changes were to all intents and purposes stationary.

Peripheral areas and the crustal forebulge zone

One of the common misconceptions regarding sea level change is that areas of the world that have and are experiencing glacio-isostatic rebound are less affected by recent sea level rise, since rising seas are offset by land uplift. The misconception arises because areas located near to and beyond the former ice sheet margin lie close to or within the zone of crustal forebulge deformation. During deglaciation, these areas experience forebulge collapse and are characterised by relative sea level curves characterised by a continuous

rise throughout the Lateglacial and Holocene (e.g. the east coast of the United States). The zones of forebulge collapse associated with each former ice sheet are hard to define, but occur within zones in the order of hundreds of kilometres beyond the respective ice sheet margins. Also, different collapsing forebulge regions may intersect with each other. For example, the North Sea and northeast Atlantic regions may have been affected by forebulge collapse associated with both the last British and Scandinavian sectors of the last Fennoscandian ice sheet.

In attempting to understand patterns of relative sea level change in such areas, one must also remember that the regions of forebulge collapse during regional deglaciation were not only affected by crustal downwarping and glacio-eustatic sea level rise, but also by the geoidal lowering of sea level due to the loss of gravitational attraction of ocean water to melting ice sheets. Each of the sea level index points that are used to construct the relative sea level curves for such areas represents the net influence of all of these three processes. For some areas a fourth element, tectonic subsidence, also needs to be considered in order to understand the shape of individual sea level curves. A good example of this is the southern North Sea region, which is both a sedimentary basin subject to crustal subsidence, but also lies within a forebulge collapse zone located south and east of the last British ice sheet. The theme of forebulge collapse is also applicable to North America where, for example, a broad zone of collapse extends southward along the Atlantic seaboard of the United States and Canada. Forebulge downwarping is particularly evident from tide gauge records between New York and Delaware and may also have affected the adjacent continental shelf.

Farfield regions

One of the most striking features of sea level curves for equatorial regions is that relative sea level reached higher than present during the mid-Holocene. For the most part, the curves that are available resemble those constructed from submerged coral data, except for a pronounced highstand during the mid-Holocene that is generally thought to have occurred between *c*.6000 and *c*.4000yr BP. From one point of view, this does not make any sense, since there is no obvious source/s for this extra ocean volume mass. After all, the last remnants of the last great ice sheets in the northern hemisphere had completely melted by *c*.7000yr BP and one might have imagined that during the millennia that followed, sea level in farfield areas would have remained in the same approximate position. One might sensibly turn to Antarctica or Greenland for the extra water from ice melt to explain the higher sea levels, but there is no scientific evidence for widespread ice melt in these areas at this time.

A possible explanation for this enigma, first discussed by Clark and then by Walcott, is that during the Holocene a low-viscosity asthenosphere enabled the overlying oceanic lithosphere to sag downwards beneath the huge volumes of water that had been returned to the world's oceans following the melting of the ice sheets. With more and more ocean 'space' available to accommodate this water, the ocean floor

experienced hydro-isostatic deformation and lowering, causing sea levels to fall across low latitudes (Fig. 10.6).

Contrast the above with areas of continental shelf where loading by ocean water is compensated by uplift across areas of landmass. This process, referred to by Jerry Mitrovica and Glenn Milne as continental levering, is still continuing today, so much so that current measured rates of sea level rise have to be adjusted to compensate for this levering effect. These two processes of continental levering and crustal forebulge collapse are both mentioned here, since they are connected in terms of the geological processes that cause them. However, as far as farfield ocean areas are concerned, sea level lowering takes place partly as a result of downwarping of the ocean floor, and also as a result of what Dick Peltier

has referred to as ocean syphoning of seawater poleward out of farfield areas towards the areas of forebulge collapse.

When one considers all of the complex processes that determine the patterns of relative sea level change across the Earth's surface, it is self-evident, as Nils-Axel Morner pointed out years ago, that there is no such thing as a global sea level curve. This realisation has taken a long time to become accepted by the scientific community. The relevance to current debates on future sea level rise is obvious. The implication is that present-day sea level is not rising everywhere at the same rate – indeed, in some areas of the world it is falling. Unfortunately, the mainstream media lags decades behind the key scientific advances in this field, and continues to adhere to the holy grail of a uniform 'global sea level' rise.

Figure 10.6 Mid-Holocene emerged notch, south coast of Vanua Levu, Savusavu, Fiji. Image credit: James Terry.

11 Present and future relative sea level change

Introduction

In this chapter an attempt is made to respond to some of the key questions that society needs to know regarding present and future trends in sea level worldwide. The lesson from the preceding chapters demonstrates that this is both a very complex subject, and that much remains unknown. Reference is made frequently to the reports of the Intergovernmental Panel on Climate Change (IPCC) and their influential observations on the nature of past, present and future trends in relative sea level. To the untrained, much of what is written in these reports is shrouded in jargon and it is often difficult to see the wood from the trees. There are five IPCC reports, the first published in 1990 and the most recent in 2014. Reading the sections of these reports that consider sea level change, one is struck by how much our knowledge has advanced over what seems often a lightning-fast 25 years during which the advent of satellite technology and its use in measuring the sea surface worldwide has led to enormous advances in our knowledge. The measurements made from the GRACE satellites in particular represent a huge advance in our understanding of ocean mass changes. Given the complexity of the topic, this chapter can only provide a broad summary of the issues surrounding the sea level rise debate. It is structured around six questions:

◆ What is the likelihood of a future collapse of the West Antarctic ice sheet and what are the consequences for sea level rise?
◆ How much does the Greenland ice melt contribute to current sea level rise and what is the likelihood of collapse of the Greenland ice sheet?
◆ What is the contribution of meltwater from mountain glaciers and ice caps to sea level rise and is this likely to change?
◆ What is the role of thermal expansion of ocean water in relative sea level change and how is this likely to change in the future?
◆ What is the view of the Intergovernmental Panel on Climate Change regarding recent rates of sea level rise?
◆ What is the effect of glacio-isostatic adjustment on current sea level trends?
◆ What is the contribution from water storage on land?

What is the likelihood of a future collapse of the West Antarctic ice sheet and what are the consequences for sea level rise?

One of the key issues in the current debate on sea level rise is whether there is any likelihood that the West Antarctic ice sheet (WAIS) would be liable to collapse under future scenarios of global warming (Fig. 11.1). The topic was brought to international prominence over 40 years ago by John Mercer, who as early as 1972 had suggested that such a scenario might occur in the

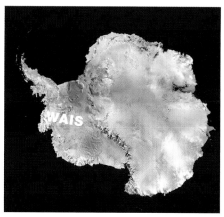

Figure 11.1 Whereas the East Antarctic ice sheet is relatively stable, the West Antarctic ice sheet (WAIS) is in many areas a marine ice sheet and much more sensitive to climate change. Image credit: NASA, Dave Pape.

future as a result of CO_2-induced global warming. This issue has been debated ever since, and formed an important part of all of the Intergovernmental Panel on Climate Change considerations since their first report in 1991. In the IPCC 2007 report, reference is made to a potentially irreversible 5m rise in sea level caused by future melting of this ice sheet. But is the WAIS showing signs of instability? On the one hand, higher global air temperatures in the future, together with a warmer Antarctic Circumpolar Ocean, may lead to increased snowfall accumulation. On the other hand, could the collapse or break-up of floating ice shelves across West Antarctica affect the rate of discharge of ice from the interior of the ice sheet, thus promoting greater instability? Also, could the flow rates of outlet glaciers and ice streams increase in the future as a result of the incursion of seawater beneath ice at the coast, and also by changes in ice ablation at the surface? In these respects it should always be remembered that much of the topography beneath the West Antarctic ice sheet is presently subject to glacio-isostatic deformation, and much of the land surface below the ice sheet is located below present sea level.

Is the West Antarctic ice sheet disintegrating? The simple answer is that we just do not know. The reason for this is that insufficient glaciological research has been undertaken to be sure how such an ice sheet is likely to behave under different scenarios of future climate change. The WAIS is a predominantly marine-based ice sheet flanked by floating ice shelves that are fed by ice flow from the interior of the ice sheet across grounding lines, i.e. the positions where the bottom of the ice becomes frozen to bedrock (cf. Fig. 7.6). If seawater comes into contact with basal ice near the grounding line, the transfer of sensible heat from water to ice can cause the grounding line position to shift inland, causing ice shelves to lose some of their mass. The consequence of this process is that ice flow in the interior accelerates gradually towards the ocean and, as it does so, the ice streams increase their velocities but become thinner. In this way, greater and greater quantities of ice are removed year after year. Much also depends on the nature of the land surface beneath the ice close to the grounding line. If the

landscape beneath the ice slopes steeply inland close to the grounding line, any small shift landward in the grounding line position can lead to the invasion of large quantities of seawater and the rapid destabilisation of the ice shelf. The importance of grounding lines and their susceptibility to shift positions is stressed here, since it is a key mechanism that can lead to the collapse of sections of any marine-based ice sheet. Should such a catastrophe happen, it would not happen instantaneously; it is more likely that rates of ice flow and ice thinning would accelerate, with collapse of sections of ice taking place over centuries rather than decades.

The greatest concern for WAIS ice sheet instability is in the Amundsen Sea area, where Pine Glacier and Thwaites glacier are particularly vulnerable (Fig. 11.2). Both have grounding lines located to the seaward of deep areas of water, and hence both are susceptible to collapse should the respective grounding retreat further. Geologists and numerical modellers are currently undertaking urgent investigations of this area in order to understand more clearly the sensitivity of both outlet glaciers to grounding line changes, and also what would be the consequences in respect of future sea level change should these sections of the WAIS begin to decay. It would not be an underestimation to state that the glaciological changes in the Amundsen Sea area may prove to be a critically important element in shaping the future course of sea level changes around the world (Fig. 11.3).

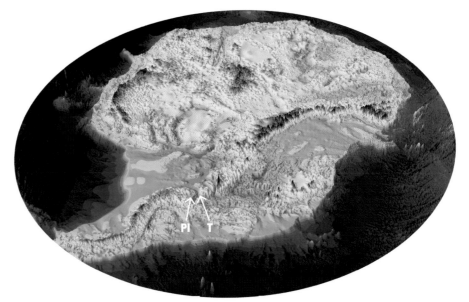

Figure 11.2 The sub-ice topography across Antarctica. The dark blue and light blue denote those areas presently located below sea level. This bedmap illustrates the extent to which the WAIS is marine-based. The location of Pine Island (PI) and Thwaites (T) outlet glaciers is also shown. Image credit: British Antarctic Survey, Peter Fretwell.

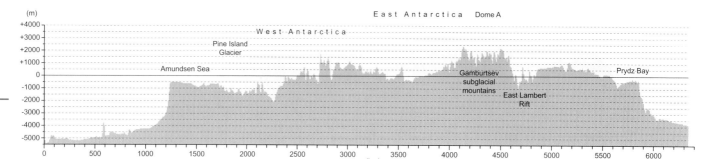

Figure 11.3 A 6000km cross-profile across Antarctica extending from the Amundsen Sea and Pine Island glacier area in the west to Prydz Bay in the east. Note that all of the sub-ice topography across West Antarctica in this profile occurs well below present sea level. Image credit: British Antarctic Survey, Peter Fretwell.

These uncertainties are illustrated in the various IPCC reports that provide a list of the most important contributors to the budget of global mean sea level change. For example, in the IPCC 2007 report the Antarctic ice sheet (East Antarctica plus WAIS) is estimated to have contributed 0.14 +/− 0.41mmyr[-1] towards a total amount of rise that took place between 1961 and 2003. In the succeeding reports for the period 1993–2003, this amount had increased to 0.21 +/− 0.35mmyr[-1]. Compared to the current average rate of rise of sea level as measured by satellite altimeter (+3.2mmyr[-1]), this value for ice loss expressed in terms of sea level equivalent is tiny.

There is one further complication – namely the geoidal sea surface around Antarctica. If the geophysicists are correct in their modelling research, the geoidal sea surface surrounding both Antarctic sheets is 'warped' upwards from the open ocean towards the ice sheet edge as a result of gravitational attraction. The zone of geoidal ocean surface deformation extends away from the ice sheet into the open ocean perhaps for distances of c.1000–1500km.

Should the West Antarctic ice sheet start to disintegrate, let's say over several centuries as some IPCC scenarios would have us believe, an important effect, in addition to glacio-isostatic rebound, would be to 'relax' the gravitational attraction of water to ice, thus causing sea level to fall across large areas of the Southern Ocean. Two of the first people to demonstrate this were Clark and Primus c.30 years ago. Their modelling demonstrated, for example, that an 'instantaneous' contribution of 1m of ice volume sea level equivalent from Antarctica to the world's oceans would result, due to reduced gravitational attraction of ocean water to the ice sheet, in a fall in relative sea level across the entire Antarctic Circumpolar Ocean, with a sea level lowering extending as far north as southern South America, southern Africa and southern Australia. Areas farther north would experience a relative sea level rise, with the greatest amounts of rise occurring across the North Pacific Ocean and central and northern parts of the Atlantic Ocean.

In such an ice melt scenario, glacio-isostatic rebound in conjunction with gravitational lowering would act together to enhance a relative sea level lowering around West Antarctica. An unknown factor is uncertainty surrounding the location and extent of the proglacial forebulge zone. Although it already exists around the Antarctic continent, its precise topography is not known. One further unknown is how much glacio-isostatic adjustment has already taken place across WAIS. The lesson learned from John Andrews in respect of the melting of the Laurentide ice sheet is that after the LGM a huge amount of glacio-isostatic rebound took place while the ice sheet remains essentially intact, yet subject to both surface ablation and surface lowering. Could these processes already be happening across West Antarctica, and do we yet know if the WAIS has started to disintegrate? The answer is that we do not know. Part of the difficulty is having insufficient data – something that the NASA airborne mapping project IceBridge is going some way to solve. Much of the current international research effort regarding the WAIS is focused on repeated satellite measurements of changes in surface elevation across the ice sheet surface to detect areas of accelerated glacier flow. Satellite remote sensing is also being used to map inferred changes in grounding line positions.

Move forward to the IPCC 2013 report, and we find some different figures quoted, in this case based partly on direct observations from tide gauges and satellites but also from modelling. These suggest that both Antarctic ice sheets (east plus WAIS) between 1993 and 2010 had contributed 0.27mm +/−11mmyr^{-1} of ice volume water equivalent to the world's oceans. Given that an ice volume sea level equivalent of c.60m of ocean water is locked up in both ice sheets, the contribution from melting ice to averaged global sea level is exceptionally small indeed. In respect of the contribution of the WAIS to sea level rise, the jury is still out. In the short term the ice volume contribution is very small. However, from a long-term perspective (decades or centuries) the risk may be considerably higher.

How much does the Greenland ice melt contribute to current sea level rise, and what is the likelihood of collapse of the Greenland ice sheet?

The Greenland ice sheet differs from the East and West Antarctic ice sheets in that it mostly lacks large numbers of tidewater outlet glaciers and ice shelves (Fig. 11.4 top). In a similar manner to East Antarctica, the terrain underneath the ice sheet is depressed due to glacio-isostatic deformation. However, whereas the land beneath West Antarctica consists of a chain of mountains, many of which occur below sea level, the effect of glacio-isostatic depression across Greenland has been

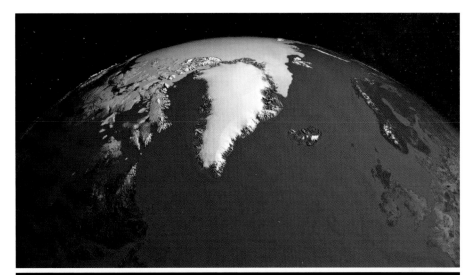

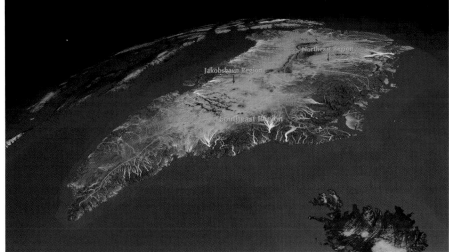

Figure 11.4 Unlike West Antarctica, the Greenland ice sheet (top) lacks major ice shelves. Instead, its mass balance is predominantly influenced by surface accumulation and ablation processes. Bedrock topography of Greenland (bottom) with the ice sheet removed. Areas below sea level are shown in brown while areas above sea level are coloured green. Yellow areas indicate regions at sea level. The blue/white flows indicate the direction and speed of the ice movement where slower-moving ice is shown as shorter blue flow lines and faster is shown as longer white flow lines. Image credit: NASA.

to create a lowered central area that is surrounded by a peripheral zone of higher topography (Fig. 11.4 bottom).

In general, the amount of ice lost in Greenland due to surface melting is similar in amount to that lost as a result of iceberg calving. This means that, in contrast to the Antarctic continent, the mass balance of the Greenland ice sheet is much more sensitive to the effects of changes in temperature and snowfall. In particular, summer temperatures across the ice sheet in recent years have been sufficiently high to cause large amounts of surface runoff of meltwater sufficient to account for around half of the annual loss of ice.

At the time of the first IPCC Report in 1990, scientists knew relatively little about the mass balance of the Greenland ice sheet, nor of the respective roles played by recent changes in air temperature and precipitation. Analysis of available precipitation and temperature data for the period 1880–1980 yielded a contribution to sea level rise equivalent to 23 +/−16mm/100 years or 0.23mmyr[-1]. Both the IPCC 1995 and 2001 reports put forward the view that the mass balance of the ice sheet had remained essentially unchanged in recent years, although it was suggested that the ice sheet may have contributed slightly to sea level rise during the first half of the twentieth century but close to zero by the end of the century. The 2007 report suggests little change, with the net contribution to sea level rise from the ice sheet between 1993 and 2003 being in the order of 0.21mmyr[-1] +/−0.07mmyr[-1]. According to the 2013 IPCC report, the contribution of Greenland ice melting to average global sea level between 1992 and 2001 is considered likely to have been between −0.02mmyr[-1] (a net gain of ice) and 0.20mmyr[-1]. However, between 2002 and 2011 this may have increased to 0.59mmyr[-1] +/−0.16mmyr[-1]. That there has indeed been a net loss of

ice to the ocean is consistent with the presence of an extensive thickness of low-salinity ocean water that occupies the area between southern Greenland and Labrador. One of the most significant results of recent research is from Eric Rignot and his team, who applied a new method to calculate an 18-year time series of recent changes in the mass balance of the Greenland ice sheet and then compared their results with those of GRACE analysis. They found both methods to be in broad agreement, and concluded that the mass balance of the ice sheet had decreased consistently since c.1999. By 2010 this loss had fallen to −350Gtyr[-1]. If the results of Rignot and his team are correct, this demonstrates scientific research running ahead of IPCC projections and, citing the authors, '...ice sheets will be the dominant contributors to sea level rise in forthcoming decades, and will likely exceed the IPCC projections for the contribution of ice sheets to sea level rise in the twenty-first century'. More recently, Anny Casenave and her team have estimated that the Greenland ice sheet contribution to global mean sea level between January 1993 and December 2015 has been in the order of 0.37mmyr[-1] +/−0.28mmyr[-1].

When ice loss from Greenland and Antarctica are combined together, the results suggest that during 2006, for example, the two ice sheets experienced a net mass loss of 475 +/−158 Gtyr[-1] equal to a global average sea level rise of +1.3 +/− 0.4mmyr[-1]. Perhaps more significantly, they concluded that the acceleration of mass loss from the two ice sheets is now three times larger than for mountain glaciers and ice caps. As in the case of the WAIS, the sea level changes that might arise are

more complicated. For example, if the rate of net mass loss across Greenland continues to increase annually, one might expect also an increase in the amount of glacio-isostatic rebound and a corresponding subsidence of proglacial forebulge areas while, as the ice sheet continues to thin and retreat, the geoidal sea surface will also relax due to weakening of the gravitational pull.

The inference from the above is that a progressive retreat and thinning of ice across Greenland would be accompanied by a fall in relative sea level at the Greenland coast. Jerry Mitrovica and Glenn Milne have opened our eyes regarding the implications of what might happen if the Greenland ice sheet were subject to widespread melting just as may have happened during marine oxygen isotope stage 5e. They point out that should such a catastrophe happen in the future, coastlines as far away as Scotland might experience no change in relative sea level since the glacio-eustatic rise in sea level would be almost entirely offset by geoidal sea surface lowering (Fig. 11.5). The counterpoint to these observations is that farfield regions in the low latitudes and across the southern hemisphere would experience a rise in relative sea level due to a glacio-eustatic contribution coupled with the absence of any appreciable gravitational effect. These predicted patterns of change are opposite to those associated with a future collapse of the WAIS, which would be accompanied by a relative sea level rise that would be greatest along coastlines in the northern hemisphere and less pronounced in the southern hemisphere!

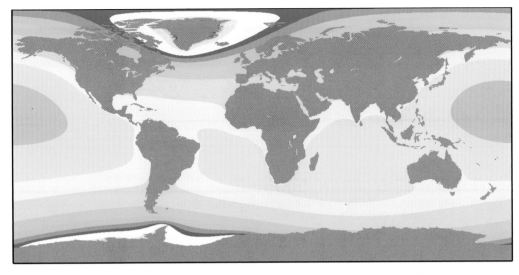

Figure 11.5 Theoretical modelling of regional sea level changes caused by annual contributions of 0.5mmyr^{-1} from both the Greenland and Antarctic ice sheets. The work, undertaken by Glenn Milne and colleagues, shows, for example, that coastlines bordering NW Europe would experience no change, and in some areas a fall in relative sea level. By contrast, farfield areas would experience a greater than average rise. Image credit: Glenn Milne.

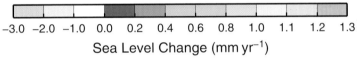

$-3.0 \quad -2.0 \quad -1.0 \quad 0.0 \quad 0.2 \quad 0.4 \quad 0.6 \quad 0.8 \quad 1.0 \quad 1.1 \quad 1.2 \quad 1.3$

Sea Level Change (mm yr^{-1})

What is the contribution of meltwater from mountain glaciers and ice caps to sea level rise, and is this likely to change?

The distribution of the world's mountain glaciers and ice caps spans a huge range of latitudes and altitudes. Each has a different source of snow precipitation and a different seasonal temperature regime. Accordingly, each responds differently to the effects of climate change. Mountain glaciers can contribute to glacio-eustatic sea level rise as a result of losses in surface mass balance or as a result of discharge of glacier ice directly into the ocean (Fig. 11.6).

Most alpine glaciers have been undergoing retreat over the last century as a result of negative surface mass balances. A precise determination of the length of time that glaciers have been in retreat is not possible, although the popular view is that retreat has been widespread since the close of the Little Ice Age at the end of the nineteenth century. Probably the most definitive early statement on the relationship between mountain glaciers and sea level rise was by George Meier in 1984. Meier calculated that over the period 1900–1961 glacier retreat equated to an average annual rise in sea level of 0.46 +/–0.26mmyr^{-1}. In 1990, the first IPCC report noted that since global air temperatures had increased by 0.35°C over the same time interval, this enabled a crude sensitivity index to be estimated whereby sea level would rise from this source alone by 1.2mmyr^{-1} +/–0.6mmyr^{-1} for every 1°C increase in average air temperature. In the 2001 IPCC report the results of several more recent studies covering the period between 1960 and 1990 showed an average equivalent sea level rise value of 0.25 +/–0.10mmyr^{-1}. By the time of publication of the IPCC 2007 report, advances in mapping and measurement of the world's glaciers and small ice caps by remote sensing led to a refinement of the ice volume sea level equivalent locked up in these ice masses. This report noted that if one included the glaciers and ice caps surrounding the Greenland ice sheet and the WAIS, but omitting those on the Antarctic Peninsula as well as those surrounding East Antarctica, the ice volume equivalent equalled a sea level contribution of 0.72 +/–0.2mm. For the period 1961–2003, it was reported that there had been an increase in ice volume loss, with the average rise attributed to this source rising to 0.50+/–0.18mmyr^{-1}, but that this value had risen to 0.93mmyr^{-1} for the period 1994–2003.

In the fifth IPCC summary, the sea level contribution of all glaciers excluding those surrounding the large

Figure 11.6 Monitoring of the Franz Josef Glacier, New Zealand, between April 2012 (top) and January 2017 (bottom) reveals significant ice retreat. Image credit: Simon Cook.

ice sheets was revised upwards. For 1901–1990 it was estimated to have been equal to a rise rate of 0.54 +/− 0.07mmyr^{-1}, i.e. similar to the above estimate for 1961–2003. For 1971–2009 the value was 0.62 +/−0.35mmyr^{-1} but for 1993–2009 it had increased to 0.76 +/− 0.37mmyr^{-1} and for 2005–2009 the value had risen to 0.83 +/−0.37mmyr^{-1}. The cited values have relatively large standard errors associated with them, but it must be remembered that not all glacial meltwater will reach the sea; some will be stored as groundwater while other water will be intercepted and stored naturally within individual river catchments as well as through man-made dams and reservoirs. The trend is very much as one might expect in that, as global average air temperatures continue to increase, more glaciers are likely to experience negative surface mass balances with greater volumes of meltwater reaching the world's oceans.

What is the role of thermal expansion of ocean water in relative sea level change and how is this likely to change in the future?

To the general public, the most common view is that climate change means global warming, and this means melting of glaciers and ice sheets which, in turn, means sea level rise. Yet as we have seen, despite the fact that the contribution of melting ice to sea level rise is significant in its own right, a key contributor to the current rise in sea level is due to the thermal expansion of ocean water – a steric change rather than a glacio-eustatic change. How can this be?

In order to answer this question, it is necessary to return to the concept of global warming, because rising air temperatures lead to rising ocean temperatures. This is because the oceans act as a gigantic heat store. As average air temperatures rise worldwide much of the heat is stored in the first few metres of the ocean surface. For example, Levitus and co-workers demonstrated from modelling that during the twentieth century there was an increase in global ocean heat content in the order of 2×10^{23} Joules, whereas the comparable value for the global atmosphere was 0.1×10^{23} Joules. The response of the ocean to heating is not, however, simple since the principal effect of heating the ocean is to decrease the density of seawater. Expressed simply, a decrease in the density of ocean water will cause a steric rise in sea level. But the density of seawater is not simply a function of water temperature; it is also a function of salinity. Not only does ocean salinity vary markedly between one area of ocean and another, but so also do sea temperatures, and these also vary with water depth. A consequence of these relationships would be, for example, that ocean water at 2°C with a salinity of 36.0 psu (Practical Salinity Unit) and heated by an additional 2°C would undergo a smaller density increase than a corresponding ocean water mass at 15°C with a salinity of 34.0 psu (Practical Salinity Unit) that also experiences the same amount of heating (Fig. 6.11). Since

ocean water density changes drive the thermal expansion process, it follows that different water masses at different latitudes will respond differently to identical water temperature increases.

In which parts of the ocean water column is most of the excess heat taken up? Clearly most excess heat will be absorbed near the surface, with decreasing amounts taken up at greater and greater depths. Oceanographers most commonly consider that the top 700m of ocean water absorbs the majority of the heat energy.

More recently, Anny Casenave and her research group have improved this estimate, using a range of datasets to show that slightly different results emerge if one not only considers changes in ocean density between 0 and 700m water depth but also of data covering the 700–2000m depth range as well as limited data for water depths greater than 2000m. Casenave showed that the effect of increased ocean warming between January 1993 and December 2015 equated to 1.20mm +/–0.23mmyr^{-1} of global mean sea level rise, and that a separate calculation for the period January 2005 until December 2014 is slightly lower at 0.97mmyr-1 +/–0.15mmyr^{-1}.

In terms of projecting how much thermal expansion of ocean water is likely to take place in the future, one must remember always that the predicted values of sea surface rise based on decreasing ocean density are dependent on accurate modelling of future changes in the temperature of the atmosphere. These predictions are, in turn, strongly influenced by modelled prediction of future CO_2. The science of thermal expansion

sea level prediction is therefore one where we can be sure of the direction that the trend is taking us, but we can be less certain about the numbers involved. In the latest IPCC report the observed contribution of thermal expansion to global mean sea level rise was estimated at 1.1 +/–0.3mmyr^{-1}.

What is the view of the Intergovernmental Panel on Climate Change regarding recent rates of sea level rise?

Probably the most significant observation made in respect of sea level change in the fourth IPCC report was the discrepancy between the observed rate of sea level rise for the period 1961–2003 (+1.8 +/– 0.5mmyr^{-1}) and the amount estimated by adding together the contributions from thermal expansion, glaciers and ice caps as well as the Greenland and Antarctic ice sheets (+1.1+/– 0.5mmyr^{-1}) (table 11.1). Multiplied by 42 years this amounts to 29.4mm of equivalent sea level, equal to approximately 250Gt of ocean water or 5.95Gtyr^{-1}, unaccounted for. A similar discrepancy was also highlighted in 2002 when Walter Munk drew attention to the enigma that for the twentieth century the addition of the thermal expansion and glacio-eustatic components contributing to sea level rise fell short of the measured rise by an amount equal to c.1.2mmyr^{-1}.

More recent satellite measurements show that the meltwater contribution from glaciers and small ice caps is increasing with time and, despite limitations associated with the quantification of meltwater volumes reaching the world's oceans, it is anticipated that the contribution to sea level rise from these

Source	1993–2010
Thermal expansion	1.1 [0.8 to 1.4]
Glaciers except in Greenland and Antarctica	0.76 [0.39 to 1.13]
Greenland ice sheet	0.33 [0.25 to 0.41]
Antarctica ice sheet	0.27 [0.16 to 0.38]
Land water storage	0.38 [0.26 to 0.49]
Total of contributions	2.8 [2.3 to 3.4]
Observed GMSL rise	3.2 [2.8 to 3.6]

Table 11.1 Observed contributions to global mean sea level rise, 1993–2010 according to IPCC 5th report (2014).

sources will continue to increase over time. Inspection of the various IPCC reports published over the last quarter of a century reveals a marked contrast between the meltwater contribution from the Greenland and Antarctic sheets (low) and glaciers and small ice caps (high). In some senses this contrast is counter-intuitive. After all, one might expect that with global warming the largest ice masses might experience the greatest amounts of ice melting. But these ice sheets are so large that not only are their surface mass balances close to equilibrium (and in some instances positive) but they also have their own distinctive weather characterised by permanent anticyclonic circulation above the ice surfaces.

The IPCC sea level change estimates for the period 1993–2010 attribute 1.1mmyr^{-1} of rise to thermal expansion of ocean water, $+0.76 \text{mmyr}^{-1}$ to glaciers except in Greenland and Antarctica, 0.33mmyr^{-1} to the Greenland ice sheet, 0.10mmyr^{-1} to glaciers in Greenland, 0.27mmyr^{-1} to the West and East Antarctic ice sheets and 0.38mmyr^{-1} to land water storage. The sum of all the contributions is equivalent to an average sea level rise worldwide of 2.84mmyr^{-1}. Yet the observed rise

from satellite altimetry is 3.2mmyr^{-1}, thus a discrepancy still exists. Recent numerical modelling, however, suggests a much higher contribution from thermal expansion for the same time interval (1.49mmyr^{-1} compared to 1.1mmyr^{-1}) than previously believed. If the same value of 0.38mmyr^{-1} is used for land water storage, one is left with the modelled results matching the sum of the observed contributions at 2.8mmyr^{-1} However, the contribution from Antarctica is omitted from these calculations due to insufficient data. Bearing this omission in mind, the modelling results continue to underestimate the observed global average sea level rise from satellite data by 0.4mmyr^{-1} (2.8mmyr^{-1} compared with 3.2mmyr^{-1}) (table 11.1).

What is the effect of glacio-isostatic adjustment on current sea level trends?

The numerical models of glacio-isostatic adjustment made by Dick Peltier and his team have enabled this component of relative sea level change to be isolated from the satellite altimetry data. Based on Peltier's ICE-4G (VM2) model, this value is -0.28mmyr^{-1}. His later ICE-5G (VM2) model gave a value

of -0.36mmyr^{-1}. Since both values are negative, this would imply that in a theoretical Earth where there is no ocean mass change and no thermal expansion, average sea level would fall by, in this case, -0.36mmyr^{-1}. This lowering occurs because although certain areas of the world continue to experience glacio-isostatic rebound, there are extensive areas of ocean floor that are sagging downwards as a result of increased water mass. Satellite altimeters measure changes in the sea surface with reference to the centre of the Earth; in the same way ocean floors are being lowered relative to the same central point. The process of lowering again involves compression of the lithosphere upon the underlying asthenosphere, where compensatory deformation also takes place. It should be noted that the value for glacio-isostatic adjustment is a global average value and cannot be simply applied to a specific coastal area.

The key point to be remembered in relation to the glacio-isostatic adjustment process is that it has the effect of making the satellite measurements of increased sea levels lower than they otherwise would be. In other words, as long as one can make an informed estimate as to what the averaged glacio-isostatic adjustment value is, this value then has to be added to the observed sea level rise measurements in order to determine the real increase in ocean mass that has taken place over a fixed period of time. Although the process has been known for many years, and indeed Peltier provided a quantitative estimate of the nature of the process over a decade ago, many are unaware of the importance of glacio-isostatic adjustment in the sea level change debate. Without this adjustment being made, it would be impossible to determine the relative contributions of the thermal expansion of ocean water and ocean mass changes to satellite altimeter measurements of sea level rise.

What is the contribution of water storage on land?

There are number of ways in which changes in water stored on land may influence sea level variations. These include the storage of water in rivers, wetlands, within groundwater aquifers and in dams and reservoirs. Water is also stored in snow as well as within permafrost across polar regions. For example, increased summer temperatures may lead to a thickening of the active layer and a release of increased volumes of ice melt into arctic rivers. Scientists have shown that major fluctuations in water storage also take place as a result of short-term changes in weather and climate. For example, it is well known that water storage on land fluctuates dramatically during El Niño and La Niña events. For example, during major El Niño events minor increases in regional sea level often take place as a result of increased rainfall across the eastern Pacific. By contrast, La Niña events are frequently associated with a slight lowering of regional sea level as a result of increased rainfall across southeast Asia.

The effects of humans on land water storage by constructing reservoirs and extracting groundwater is not insignificant. The construction of large reservoirs serves to counteract or slow down rates of sea level rise. Many have argued that without reservoir construction since the 1930s average sea level would have been $c.30$cm (300mm) higher than it is at present. However, the building of large dams has slowed in recent decades, implying that more runoff is reaching the sea. The pumping of groundwater has had the opposite effect. At first sight one might imagine that groundwater pumping would contribute to a fall in sea level. However, most large-scale groundwater extraction is used for irrigation, drinking water, and by industry. In irrigation processes, for example, stream runoff plus evaporation and precipitation mean that much of the extracted groundwater ultimately ends up in the world's oceans. The IPCC estimate is that between 1971 and 2010 the combined effects of reservoir impoundment and groundwater pumping equated to a sea level rise contribution of +0.12mmyr^{-1}, but for the period 1993–2010 this had increased to +0.38mmyr^{-1}. In a recent study, Anny Casenave has increased this estimate slightly for the time interval 1992–2013 to +0.45mmyr^{-1} +/– 0.16mmyr^{-1}.

12 Understanding sea level change

Much of the discussion in previous chapters has focused on patterns of relative sea level change that have taken place over exceptionally long time-scales, with much consideration given to the nature of relative sea levels associated with the melting of the last great ice sheets. Yet as we have seen, some of the changes that took place over 20,000 years ago have an immediate relevance to our understanding of relative sea level changes that are taking place today. This places the processes of glacio-isostatic adjustment at the centre of this discussion. Clearly, if we are to understand contemporary satellite altimeter measurements that are showing small, yet perceptible, increases in sea level, we need to know the rates at which ocean basins have sagged downwards during the Holocene in order to accommodate increased ocean mass, and how this applies to present day ocean areas. Equally, we need to have accurate quantitative data on the way that continental shelves and formerly glaciated landscapes continue to experience glacio-isostatic adjustment (Fig. 12.1).

Given these immense complexities, it is a daunting task to try and absorb and interpret the large range of estimates of the various components of global average sea level rise. It is a challenge to make simple comparisons between different datasets, owing to each of them having been collected over different time intervals and making use of widely different methods. However, when a comparison is made between the different estimates of the components contributing to recent sea level rise, one cannot but be struck by how markedly some of the estimates have changed over the last c.20 years. At the close of the twentieth century, for example, a widely held view was that global sea level was rising on average by between c. +1.8 and +1.9mmyr^{-1} including a contribution from glaciers and ice caps of c.+0.3mmyr^{-1}. The melt contribution from ice in Antarctica and Greenland ice sheets was universally thought to be low, in the order of +0.1mmyr^{-1} while the ocean thermal expansion component was considered to be contributing c.+0.6mmyr^{-1} to the observed sea level rise. By contrast, significant quantities of meltwater were considered as being stored terrestrially rather than entering the oceans directly. This value was thought to be in the range of somewhere between –0.3 and –0.9mmyr^{-1} ice

Figure 12.1 Coastal archaeological remains in the Shetland Isles. But are rising sea levels in this area due to crustal subsidence or increased global ocean volume and mass? Image credit: Tom Dawson.

volume sea level equivalent. Lastly, the component of glacio-isostatic adjustment was considered to be acting to lower the position of global mean sea level by $c.-0.5$mmyr^{-1}. However one made the calculations, Munk's enigma remained unsolved, with the sum total of estimated contributions to global mean sea level rise being significantly less than the observed rise based mainly on tide gauge data.

At present (2018), both the (satellite) observations and the component estimates are radically different. The NASA sea level portal for March 2018 states that the most recent satellite altimetry measurements indicate a global mean sea level rise rate of $+3.2$mmyr^{-1} $+/-0.4$mmyr^{-1}, of which $+1.8$mmyr$+/-0.3$mmyr^{-1} is derived from ocean mass (glacio-eustatic) changes and $+0.8$mmyr^{-1} $+/-0.2$mmyr^{-1} is attributable to steric changes. But the gap between the satellite sea level observations and the sum of the respective contributions still does not equate. Recently Anny Casenave and her research group maintained that closure may be possible. When thermal expansion effects are included for global ocean water deeper than 700m, the steric contribution rises to $+1.20$mmyr^{-1} sufficient to bring the total, when an increase of ocean mass ($+1.8$mmyr^{-1}) is included, to $+3.00$mmyr^{-1}.

If it is indeed possible to isolate, with sufficient confidence, the relative contributions of the steric and glacio-eustatic contributions to current sea level rise, it is exceptionally difficult to quantify how changing ice and water loads across the surface of the Earth will influence changes in the geoidal sea surface. Changes in the mass balance of ice sheets, as well as changes in the distribution of mass between ice sheets and oceans, can induce gravitational changes that alter sea surface topography. We have seen how accelerated ice melt in Antarctica can induce disproportionate sea level rise in the northern hemisphere and a lowering of the sea surface around Antarctica. We have also noted how similar processes in respect of Greenland ice sheet melting could lead to preferential sea level rise across low latitudes and in the southern hemisphere, yet induce an element of sea surface lowering across the North Atlantic.

In respect of climate change predictions, a range of different scenarios exist, with most suggesting that average air temperatures will increase by between 2°C and 4°C by the end of this century. If this happens, there will be an undoubted increase in oceanic thermal expansion. Apart from uncertainties associated with the various computer models of the Earth's future climate, there may also be as yet unknown natural changes that may also alter the Earth's climate (e.g. major volcanic eruptions, changes in solar radiation, disruption to the global oceanic thermohaline conveyor, etc.). The way in which future changes affect sea level change depends upon which climate change scenario takes place – in all probability the changes that will take place are not what we expect at the present time. If global average air temperatures continue to rise, the view is inescapable that thermal expansion of ocean water will continue as an important component of sea level rise, but how much will take place at different depths in the oceans, and where? Not only that, but the increase buoyancy associated with ocean water of a slightly lower density will render the formation of North Atlantic and Antarctic Deep Water more difficult to accomplish – thus in this way, thermal expansion of ocean water may induce climate change due to the way that it affects patterns of ocean circulation.

Global warming also means that we will continue to witness surface mass balance losses across the majority of the world's glaciers and small ice caps. Ever since the early estimates made of glacier surface mass balance loss by George Meier in 1984, all estimates since then have followed a trajectory of increased ice losses caused by negative surface mass balances. Scientists are still some way off working out what proportion of ice melt from this source is intercepted on its way to the sea, but there can be little doubt that as long as air temperatures continue to rise, such losses are irreversible.

For the Greenland ice sheet and its counterparts in West and East Antarctica, the messages are much less clear. Since the first IPCC report in 1990 a distinctive feature of all documents has been the relatively small contributions of ice melt to ocean mass increase from all three major ice sheets. This is a great irony, since the vast majority of the world's ice is stored in these three regions, yet it has not been melting significantly. One view is that both the West and East Antarctic ice sheets may experience a positive mass balance in a warmer world as a result of increased snow precipitation. Despite the sense of security that one might derive from this observation, there are worrying signs. The greatest concern focuses around sections of the West Antarctic ice sheet, in particular the Amundsen Sea area, which are marine-based and potentially unstable. The key issue is the positions of the various grounding lines and how these positions are likely to change in the future. Specifically, what happens to sections of ice sheet where destabilisation of grounding lines leads to incursions of marine water beneath glacier ice? That said, once an ice sheet thins and retreats sufficiently to become entirely land-based, it can stabilise.

Ever since the classic studies of John Hollin, John Mercer, Nils Reeh and others it has been speculated that either the West Antarctic ice sheet or the southern half of the Greenland ice sheet may have melted during isotope substage 5e, leading to a relative sea level rise around the world of somewhere between +1 and +5m. Is this something that we might expect to happen during the twenty-first century? Or indeed, might it be the case that the timescale needed to melt such huge volumes of ice would be over centuries rather than decades?

Lessons can be learned from the sea level curves for the Late Pleistocene and Holocene. In particular, the submerged coral reef sea level records point to a number of meltwater pulses during the last deglaciation having taken place, with each appearing to have been associated with rises in globally averaged sea level in the order of 30–40cm over $c.$400–500 years and caused by accelerations in ice sheet melting. Today our attention needs to be directed towards the stability or otherwise of the Greenland and West Antarctic sheets, and to a lesser extent the

East Antarctic ice sheet. One certainty is that satellite altimetry continues to routinely measure and map the changing surfaces of these ice sheets. If there is an acceleration of surface thinning, we are sure to know relatively quickly.

One area of sea level change research that the IPCC does not discuss in any detail is in respect of glacio-isostatic adjustment initiated $c.$20,000 years ago and still continuing. These adjustments include glacio-isostatic rebound across formerly glaciated areas, crustal forebulge collapse across areas surrounding the former ice sheets, the flexure (levering) of continental shelves and the subsidence of ocean floors. We must also be aware of the role of gravitational attraction of ocean water to ice sheets. As Jerry Mitrovica has described, if, for example, the Greenland ice sheet were to melt tomorrow, relative sea level might fall across an area of ocean surface $c.$1000–1500km wide surrounding the ice sheet. At the same time the fall in relative sea level caused by this process would be countered by a glacio-eustatic rise due to ice melt. For an area such as Scotland, the two processes would be likely to cancel each other out, resulting in no change in relative sea level! By contrast, the effects of such ice melting would be felt most greatly in farfield ocean areas located distant from the former ice sheets.

In these first decades of the twenty-first century we are beginning to understand much better the key processes that contribute to sea level rise – and yet there is still much that we do not know. Our discussion of sea level change past, present and future, has drawn a multitude of scientific disciplines each contributing in

specific ways to our understanding of the subject. For example, in this book reference has been made repeatedly to issues such as geoidal sea surface change, Earth radius changes, transfer of ice and ocean water mass across the Earth's surface, gravitational changes, glacio-isostatic adjustment, the role of the asthenosphere, etc. But to a student of climate change many of these issues may be unintelligible. Similarly, discussions of the construction of shoreline height–distance diagrams, the stratigraphy of isolation basins or grounding-line dynamics of marine-based ice sheets may be unintelligible to the geophysicist. We depend on the results of geophysical modelling by Dick Peltier, Kurt Lambeck, Mitrovica, Milne and others to inform us of likely future changes in relative sea level across the globe and to give us a different perspective on how and why sea level changes around the globe.

The science of sea level change is multidisciplinary in nature. Yet none of us, or very few, are scientific experts in all the key fields. This means that when we start to try and predict what future changes in relative sea level may lie ahead, we struggle. Whilst this introduction to sea level change may not provide all the answers that we are looking for, it will at least give comfort in the knowledge that the view from the beach and the view from space are different pictures of the same image. In many scientific publications on the issue of sea level rise, most attention is focused on the numbers and rates of change, etc. Much less attention is focused on the environmental processes that drive the observed patterns of sea level change. The key lesson from this book is that without a proper understanding of the processes responsible for sea level changes that have taken place in the past and at present, numerical projections of future sea level change mean very little.

We conclude our discussion with a graph that depicts several future trends in sea level predicted by the IPCC (Fig. 12.2). The diagram is one of many such graphs that can be found through various media outlets. Governments take note of such predictions and use them to prepare national policies that address the issue of how one ought to adapt to the threat of future rises in sea level. In this book, however, we have

seen that the term 'global sea level' means very little. Patterns of relative sea level change vary widely between one region and another. The Potzdam 'gravity potato' depiction of the Earth's geoidal sea surface tells us that very clearly. It should always be remembered that the IPCC lines depicting future trends in sea level are generated as a result of numerical modelling. For example, various climate change models are used to predict how much thermal expansion of ocean water would take place in the future given different scenarios of increased global air temperatures, and estimates of how much extra heat is likely to be

stored in the ocean. Other models have attempted to estimate how much melting of inland glaciers and small ice caps is likely to take place under different scenarios of global warming.

The two processes described above (thermal expansion of ocean water and the melting of inland glaciers and small ice caps) probably represent the easier parts of the analysis of past, present and future sea level changes. Attempts to quantify the remaining environmental processes that contribute to changing sea level are altogether much more complex. For example, we are only beginning to understand the relationships between mass balance changes of the ice sheets in Greenland and Antarctica and patterns of climate change. It is no coincidence that the IPCC leave future sea level rise contributions from these sources as 'blank' in their science documents. Even those brave enough to predict future amounts of ocean mass increase derived from these sources place such huge standard error values around the numbers as to render them effectively meaningless. Yet there are danger signs. Probably the most important is how to understand the changing ice dynamics currently taking place across the Amundsen Sea area of West Antarctica. Our assessment of future change depends very much on how we interpret the responses of local ice sheet grounding lines to changes in oceanographic processes and regional ice dynamics. We have also seen how melting ice sheets result in changes in the gravitational attraction between ocean water and ice, and how such changes may lead to regional changes that can affect the majority of the world's ocean areas.

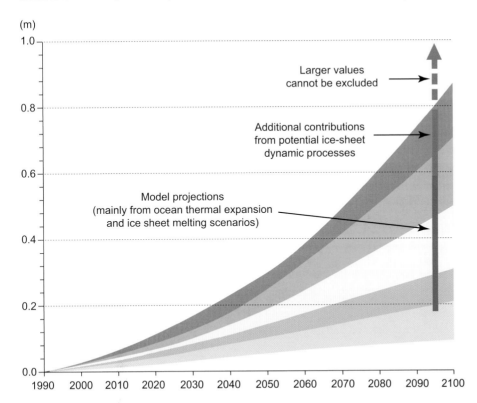

Figure 12.2 Plot of future IPCC sea level rise scenarios with qualitative comments on probabilities of specific changes taking place. Image credit: adapted from CSIRO.

In summary, we have learned that the subject of sea level change is exceptionally complex. Two key issues need to be remembered. The first is that future predictions of sea level rise, such as that shown in the diagram above, should be examined in a highly critical manner. After all, without a proper science-based understanding of the processes responsible for sea level changes that have taken place in the past and into the present, numerical projections of future sea level change do not mean very much. The second issue is that there is a clear need to be very cautious in our approach before accepting some of the more dramatic predictions of future sea level change that we encounter every day in media outlets. In many respects we are only at the start of the journey of understanding sea level change, with our 'potato-shaped' geoidal sea surface having many new secrets to reveal.

Glossary

A

ablation [8]: the process by which glaciers and ice sheets lose mass. Ablation can be caused by surface melting, run-off, iceberg calving and by basal melting.

abrasion [11]: in sea level research, abrasion refers to the mechanical erosion of coastal rock surfaces due to friction exerted by moving debris and smaller particles.

accumulation [8]: the process by which glaciers and ice sheets gain mass. This process mainly takes place as a result of snow precipitation.

active layer [79]: ground areas in permafrost environments that thaw during summer and freeze again the following autumn.

alluvial fan [13]: an apron-shaped deposit of sediment produced at the base of hillslopes by stream action.

anticyclonic circulation [78]: atmospheric circulation around a central area of high atmospheric pressure that is clockwise in the northern hemisphere and anti-clockwise across the southern hemisphere.

apparent age of seawater [28]: present-day seawater and living marine organisms have an apparent age as a result of the circulation of carbon in natural systems. The values for apparent age vary regionally from c.200 years to over 2500 years depending on the histories of the respective water masses.

ARGO [8]: a global array of floats drifting in the world's oceans that routinely measure the salinity and temperature of the upper c.2000m of ocean water.

astronomical theory of ice ages [20]: generally attributed to Milutin Milankovich, who argued that the timing of ice ages during the Pleistocene was related to long-term changes in the nature of the Earth's orbit around the Sun.

atoll [14]: a ring-shaped coral reef surrounding a central lagoon.

B

benchmark [5]: an item used to mark a point as a reference altitude linked to a specific datum. Individual bench marks may be represented by chiselled marks, metal bolts or discs set in rock, building stone or concrete.

Beringia [55]: sometimes referred to as the Bering Sea Land Bridge, which formerly connected Alaska and north-eastern Siberia.

bio-erosion [11]: the erosion of carbonate rocks as a result of grazing and boring organisms within a salt water environment.

biostratigraphy [16]: a sub-discipline of stratigraphy involving the study of fossils and their use in dating sediment and rock strata.

bioturbation [45]: in studies of ocean floor sediment cores, bioturbation refers to the disturbance, or churning, of sediment as a result of organisms burrowing.

brackish fauna [16]: fauna that live in water that has more salinity than freshwater but less salinity than seawater.

C

Catastrophic Rise Events (CREs) [57]: considered synonymous with meltwater pulses, these are periods of time during the melting of the last great ice sheets when relative sea level rose at exceptionally rapid rates.

CLIMAP [51]: the 'Climate: Long range Investigation, Mapping and Prediction' project was a major international research exercise that took place during the 1970s with the objective of producing maps of the Earth's climate and vegetation that existed during the last glacial maximum.

climate change [1]: represents a change in patterns of weather that continue for an extended period of time, usually greater than one month, but can also occur over millions of years.

cold-based ice sheet [47]: an ice sheet where the basal ice occurs below pressure melting point. As a result, much of the basal ice is frozen to the underlying ground surface.

continental levering [69]: a process whereby the sagging of the ocean floor in farfield areas as a result of increased water load is accompanied by lowering of continental shelf areas and uplift of adjacent continental land masses.

continental shelf [41]: broad, relatively shallow submarine landmass representing the edge of a continental land mass. Most broad continental shelf areas occur as passive trailing plate margins. They are separated from abyssal ocean areas by areas of continental slope.

Cordilleran ice sheet [64]: an ice sheet that during the last glacial maximum was located over the present area of the western United States and Canada. The Cordilleran ice sheet merged and coalesced with the much larger Laurentide ice sheet to the east.

co-seismic [37]: refers to vertical patterns of land movements that take place broadly simultaneously during an earthquake.

crust [30]: the Earth's crust refers to its upper hard outer layer that rests upon the mantle.

D

datum [5]: a reference level from which other elevations can be measured, either as a result of ground surveying or by satellite measurements.

delta [25]: an area of sediment deposition at the mouth of a large river where it enters the sea. Deltas represent coastal areas where the delivery of sediment by fluvial activity exceeds the rate at which sediment is removed by coastal currents.

dendrochronology [29]: the scientific method of dating tree rings according to a calendar year timescale.

diatoms [17]: a major group of unicellular micro-algae, and among the most common forms of phytoplankton. Different diatom species inhabit marine, brackish and freshwater environments.

E

East Antarctic ice sheet [47]: in contrast to the West Antarctic ice sheet, this ice sheet is predominantly land-based and rests mostly on bedrock, although much of the bedrock areas rest below sea level.

elasticity of the Earth [30]: a capability of material within the Earth to respond to stresses by stretching as far as the elastic limit.

El Niño [42]: the term 'El Niño' refers to a warming of the ocean surface across the central and eastern equatorial Pacific that results in disruption of atmospheric and oceanic circulation. Typically an El Niño event occurs every few years and has a duration of several months. Frequently such events are associated with increased rainfall across western areas of South and Central America and dry conditions across Australasia, while other effects are more global in nature.

epeirogenic movements [35]: deformations of continental landmasses involving broad yet gentle upwarping or depression of extensive crustal areas.

equipotential ocean surface [2]: the global ocean surface is in gravitational equilibrium with the Earth's gravitational field. Spatial variations of the equipotential ocean surface reflect regional differences in density, mostly within the Earth's mantle. The equipotential ocean surface is the same as the Earth's geoidal sea surface.

eustasy [33]: a state of equilibrium in the shape and level of the surface of the world's oceans.

eustatic sea level changes [33]: variations in the absolute volume of water in the world's oceans.

F

farfield ocean areas [54]: regions of the world's oceans located most distant from the last great ice sheets that covered much of the northern hemisphere during the last glacial maximum.

faunal extinctions [54]: worldwide death and disappearance of diverse groups of animals over a specific time interval in the Earth's history.

Fennoscandian ice sheet [9]: a large ice sheet that existed during the last glacial maximum that extended from the British Isles in the west to European Russia in the east. This ice sheet is also referred to as the Eurasian ice sheet.

fetch [12]: the maximum length of open water across which waves can travel to reach a section of coastline.

foraminifera [16]: single-celled organisms with shells that mostly live within the world's oceans. Whereas a small number of species are planktonic and float in water, the vast majority of species live on the ocean floor (benthonic species).

forebulge [40]: an upwards bulging of areas of lithosphere surrounding an area that has been subject to crustal depression. Forebulge areas typically surround large ice sheets and are uplifted in the order of a few per cent compared to the crustal depression beneath an ice sheet. Forebulging will continue until the ice sheet load is in equilibrium with the underlying crust and mantle.

forebulge collapse [68]: the collapse of a forebulge occurs once an ice sheet starts to melt, and may continue for 10–20kyr after the ice sheet has disappeared. Like the process of forebulging, forebulge collapse is strongly influenced by the viscosity profile within the mantle.

G

geoidal sea surface [6]: the surface of uniform gravitational potential corresponding to mean sea level as measured across the global oceans. The Earth's geoid is sometimes referred to as the **equipotential ocean surface**. The geoidal sea surface does not include the effects of tides, waves and currents.

geoidal eustasy (geoidal eustatic changes) [33]: change in the distribution of ocean water caused by the effects of gravity.

Gigatonne (Gt) [8]: one billion metric tonnes. One metric tonne is 1000kg. One gigatonne (Gt) of water has a volume of one billion cubic metres. Note that one gigatonne of ice has a larger volume than one gigatonne of water, but when it melts it will still have a volume of one cubic km. Note also that since the area of the world's oceans is c.361 million km^2, 361 gigatonnes of water will raise *average* sea level across the oceans by 1mm.

glacial eustasy (glacial eustatic changes) [33]: changes in the total volume of water stored in the world's oceans due to changes in the total volume of ice stored in the world's glaciers and ice sheets.

glacial isostasy [33]: the response of the Earth's surface to the loading and unloading associated with the growth and decay of ice sheets.

glacio-isostatic adjustment [51]: the ongoing movement of land as a result of the melting of the last great ice sheets c.20,000 years ago. Material in the Earth's mantle is still moving from beneath the oceans towards areas formerly covered by ice. As a result some areas are rising and some ocean floors are sinking relative to the centre of the Earth.

glacio-isostatic rebound [19]: the vertical rebound of the lithosphere across land areas previously covered by ice sheets. Glacio-isostatic rebound thus represents a component part of glacio-isostatic adjustment.

glacial lake Agassiz-Ojibway [59]: one of the world's largest ice-dammed lakes that was impounded against the southern margin of the Laurentide ice sheet, and which is believed to have emptied catastrophically in conjunction with ice sheet disintegration over Hudson Bay c.8,300 years ago.

global warming [1]: the observed rise in average global atmospheric and ocean temperatures over the last century. As such it should be distinguished from climate change, which has been ongoing throughout the Earth's history.

GRACE [7]: the Gravity Recovery and Climate Experiment is a joint mission of NASA and the German Aerospace Centre.

gravitational attraction [34]: all objects attract each other with the force of gravitational attraction. This force is directly proportional to the mass of both objects that are interacting with each other, but it is also inversely proportional to the square of the distance of separation between the two objects. For example, if the mass of an ice sheet is halved, then the gravitational force of attraction between it and a surrounding ocean is also halved.

gravitational field strength [7]: the strength of the force that attracts objects with mass towards each other. On Earth this is equal to 10 Newtons (N) per kilogram.

gravity field [7]: the Earth's gravity field refers to spatial variations in the acceleration of objects as a result of variations in the distribution of mass within the Earth.

grounding line [48]: a theoretical line where grounded ice from an outlet glacier or ice shelf in contact with the sea reaches the point at which it starts to float. Ice located seaward of the grounding is not only subject to sensible heating from the underlying seawater but the ice itself is subject to flexure as a result of tidal changes.

H

holocene [25]: the period of time that started at the close of the last period of general glaciation, c.11,700 calendar years ago until present.

hotspots [35]: volcanic regions fed by

plumes of mantle material from adjacent mantle areas. The location of hotspot areas is usually independent of tectonic plate boundaries (e.g. the Hawaiian island chain).

hydraulic impact [11]: in coastal environments, hydraulic impact refers to erosion caused by the movement of seawater against rock surfaces.

hydro-isostasy [33]: the isostatic adjustment of the shape of the surface of the earth caused by the addition or removal of water to the world's oceans.

hydrosere [25]: a plant succession that takes place in an area of freshwater and which continues until the area dries out naturally and eventually becomes a woodland.

I

iceberg calving [74]: the fragmentation and collapse of sections of a glacier margin. The term is normally used to refer to glaciers that terminate in the sea or in lakes.

ice cap [8]: an ice mass that covers less than 50,000km^2.

ice-moulding [11]: distinctive smoothed landform shapes produced by processes of subglacial erosion and deposition.

ice sheet [2]: a large dome-shaped ice mass up to several kilometres in thickness and covering an area of at least 50,000km^2.

ice sheet profiles [46]: the cross-sectional profile across an ice sheet. In the case of former ice sheets where no nunataks protruded above the ice surface, idealised ice-sheet cross profiles are constructed theoretically on the basis of well-established mathematical models of ice flow.

ice shelf [71]: floating sheets of ice derived from tributary glaciers. Ice shelves are often associated with a back pressure that limits any inland acceleration of the tributary glacier, hindering its retreat if the ice sheet is subject to local disintegration.

ice stream [71]: area of relatively rapid ice flow within an ice sheet. The vast majority of ice discharged from, for example, the Antarctic ice sheet takes place through ice streams.

indicative meaning [23]: in sea level change research, this refers to the altitude above or below a datum level at which an indicator of former sea level change was deposited.

interglacial [9]: a period of relative warmth separating cold glacial intervals. The present interglacial, known as the Holocene, has lasted c.11,700 years. A large number of interglacials have been identified from marine sediment cores, with each identified as an odd-numbered isotopic stage.

Intergovernmental Panel on Climate Change (IPCC) [51]: the IPCC was jointly created by the World Meteorological Organisation and the United Nations Environment Programme in 1988. The IPCC has always had the mission statement of a) assessing the scientific information related to the issue of climate change and evaluating its environmental and socio-economic consequences and b) to formulate realistic response strategies for the management of climate change.

inter-seismic [55]: refers to vertical patterns of land movements across a given area that take place between separate periods of earthquake activity. Such vertical land movements essentially represent a readjustment of the lithosphere during the millennia that follow a major earthquake.

interstadial [56]: a relatively short-lived period of climate warming during a glacial episode.

island arc [35]: a chain of active and dormant volcanic islands located at a tectonic plate margin and commonly flanked by a deep ocean trench.

isobase [27]: a line drawn on a map that depicts a former sea level position. In sea level change research it is typically used to depict former shoreline positions across an area that has experienced crustal rebound as a result of the melting of former ice sheets.

isolation basin [23]: a terrestrial basin that has, at some stage in the past, been submerged below sea level. A fall in relative sea level results in the isolation of the basin from the sea and to a shift towards the accumulation of freshwater sediments.

isostasy [33]: the state of 'equilibrium' caused by the Earth's lithosphere that essentially 'floats' on the underlying asthenosphere. Deformation of the lithosphere due to the loading and unloading of ice sheets is referred to as *glacio-isostasy*. Isostatic loading and unloading by water on ocean floors and continental shelves due to relative sea level change is referred to as *hydro-isostasy*.

isotherm [31]: a line on a map showing points of equal temperature.

isotope [21]: the isotopes of an element are atoms that have the same number of protons in the nucleus but a different number of neutrons and therefore a different atomic mass.

isotopic fractionation [28]: a process that influences the relative abundance of different isotopes.

J

jokulhlaup (plural jokulhlaupe) [60]: a catastrophic river flood caused by the sudden emptying of an ice-dammed lake.

Joule (J) [77]: one Joule equals the energy transferred to an object when a force of one Newton acts on the object moving it through a distance of one metre.

L

La Niña [79]: La Niña represents ocean atmosphere fluctuations across the Pacific associated with cooler ocean temperature across the equatorial Pacific. Typically La Niña conditions are associated with dry conditions across western areas of South and Central America and increased rainfall across Australasia. Considered together, La Niña and El Niño constitute the El Niño Southern Oscillation (ENSO).

land bridge [3]: a connecting area of land that joins continents or between parts of continents or between islands, frequently linked to past changes in relative sea level. Land bridges can result from relative marine regressions during ice ages, when areas of continental shelf become exposed. They can also be created by plate tectonic processes, or in some cases as a result of differential glacio-isostatic rebound.

Last Glacial Maximum (LGM) [9]: the oxygen isotope records from deep ocean sediment cores, as well as from ice cores, indicate the maximum volumes of the last great ice sheets occurred during isotope stage 2. It is currently thought that the LGM in the North Atlantic occurred c.21–19,000 radiocarbon years ago, approximately equivalent to 28–27,000 calendar years.

last interglacial [44]: the period of warmth corresponding to oxygen isotope substage 5e, generally considered to have occurred between c.128,000 and 117,000 years ago.

lateglacial interval [28]: the last glacial–interglacial transition, known in Europe as the Late Weichselian or Late Devensian Lateglacial, marks the boundary between marine oxygen isotope stages 1 and 2. This period also corresponds with the transition between the Pleistocene and Holocene epochs.

Late Pleistocene [49]: the latter part of the Pleistocene Epoch extending from the last

(Eemian) interglacial and ending at the close of the Younger Dryas period of cold climate. It is succeeded by the present (Holocene) interglacial.

Laurentide ice sheet [9]: an enormous ice sheet that covered most of Canada and a large area of the northern United States during the last glacial maximum. To the west this ice sheet merged with the Cordilleran ice sheet.

levelling [5]: a method of surveying used to determine the altitude of a given point with reference to a chosen datum. Levelling is undertaken using an optical levelling instrument mounted on a tripod together with a metric staff.

lithology [11]: the lithology of sediment exposure describes its physical characteristics (e.g. texture, grain size, colour, composition).

lithosphere [30]: the rigid outer layer of the Earth consisting of the crust and the mantle to a depth of *c*.100km.

lithostratigraphy [23]: a sub-discipline of stratigraphy that refers to the study of sediment and rock strata.

Little Ice Age [75]: the Little Ice Age represents a period of climate cooling that took place after the Medieval Warm Period. It is unclear when the Little Ice Age started and ended. The start is generally regarded as having taken place sometime as early as the 1300s and to have ended at the close of the nineteenth century. The Little Ice Age was not a period of major ice accumulation. Instead it appears to have been a period of modest climate cooling that was most marked across the northern hemisphere.

loess [20]: aeolian sediment produced as a result of the accumulation of wind-blown silt.

M

magma chamber [37]: an area where molten rock occurs beneath a volcano and/or hotspot.

mantle [2]: the layer of the Earth between the crust and the outer core.

marine-based ice sheet [71]: an ice sheet where much of the bed is located below sea level.

marine macrofossils [17]: preserved marine organic remains visible to the naked eye.

marine microfossils [16]: preserved marine organic remains normally only visible with a microscope.

marine ostracods [16]: a sub-class of small crustaceans that live in marine environments.

marine oxygen isotope chronologies [21]: measured changes over time, derived from ocean floor sediment cores, in the ratio of the concentration of the heavy isotope of oxygen (^{18}O) with the lighter isotope (^{16}O) relative to that of a standard. Such values are typically measured from sampled foraminifera recovered from the sediment cores.

marine terrace [13]: a relatively flat or sub-horizontal coastal surface of marine origin. Such terraces are typically composed of unconsolidated sediment. In popular literature, marine terraces are sometimes described as raised beaches. Shore platforms are often mistakenly described as marine terraces.

mass balance [34]: the mass balance of an ice sheet or glacier is the net balance between mass lost by melting and calving and mass gained as a result of snowfall.

mass spectrometry [29]: an analytical technique that subjects chemical species to ionisation and sorts ions on the basis of their mass to charge ratios.

meltwater pulses [57]: periods of exceptionally rapid glacio-eustatic rises in sea level associated with the melting of the last great ice sheets. Specific meltwater pulses are sometimes described as **Catastrophic Rise Events (CREs)**.

mesosphere [30]: the part of the Earth's mantle located below the lithosphere and asthenosphere but above the outer core.

mollusca [17]: the second largest phylum (taxonomic category) of invertebrate animals. In marine environments marine mollusca are the largest phylum, constituting nearly a quarter of all marine organisms.

moraine [35]: a landform, or accumulation of glacial debris, produced at or close to the margin of a glacier or ice sheet.

Munk's enigma [81]: in 2002 Walter Munk pointed out an anomaly in the estimates that had been made of the average global sea level rise that had taken place during the twentieth century. He argued that the sum of the estimated contributions to sea level rise over this period was significantly less than the measured change.

N

NASA [7]: National Aeronautics and Space Administration.

Newtonian fluid [31]: a fluid in which the viscous stresses are linearly proportional to the rate at which the material in question is subject to deformation over time.

nunatak [35]: a ridge, peak or mountain protruding above the surface of an ice sheet, ice cap or glacier.

O

ocean syphoning [69]: the movement of ocean water from equatorial and other **far-field** areas into ocean areas experiencing *forebulge collapse*.

oceanic thermohaline conveyor [81]: see **thermohaline ocean circulation**.

oxygen isotope stage [44]: the stages in the marine oxygen isotope stratigraphic record. Odd-numbered stages refer to glacial periods while even-numbered stages describe interglacial periods. Thus stage 1 corresponds to the present Holocene interglacial and stage 5 equates with the last interglacial period. Oxygen isotope stages are sometimes subdivided into sub-stages. Thus isotope substage 5e is generally regarded as equivalent to the peak of the last interglacial at *c*.125,000 years ago. Oxygen isotope stage 2 corresponds with the last glacial period, stage 4 with an older period of glaciation and stage 6 with the second last major period of glaciation on Earth.

oxygen isotope stratigraphy [45]: the ratio of the concentration of the heavy isotope of oxygen (^{18}O) with the lighter isotope (^{16}O) relative to that of a standard. Such values are typically measured from sampled foraminifera recovered from deep ocean sediment cores.

P

paradigm shift [20]: refers in science to a fundamental change in the basic concepts and thinking within a scientific discipline.

Pascal-second (Pa s) [31]: a measure of dynamic viscosity. It is expressed in kg/m/s. Thus 1 Pascal-second equals 10 grams per centimetre-second.

permafrost [79]: areas of terrain that experience an annual negative heat budget deficit. Permafrost exists across *c*.24% of land areas in the northern hemisphere.

permanent anticyclone [47]: large ice sheets tend to generate their own weather systems. A key process is the cooling of the overlying air by the ice surface. Under such circumstances, the overlying air is forced to sink, thus increasing the average air pressure at the ground surface and thereby enabling the development of a permanent anticyclone.

Pleistocene Epoch [30]: the period of time that started *c*.2.7Myr ago characterised by a succession of episodes of glaciation. The Pleistocene period was followed by the Holocene Epoch. Both should be distinguished from geological eras, the most recent of which is the Quaternary Era.

poise [31]: a unit of dynamic viscosity.

proglacial crustal forebulge [72]: area of terrain (both above and below contemporary sea level) that immediately surrounds an area that was once glaciated. Forebulge areas occur due to the squeezing outwards of sub-crustal material from beneath large ice sheets in order to counter-balance the (glacio-isostatic) depression of the lithosphere beneath an ice-covered region. This process is associated with a local increase in the Earth's radius. Following the melting and disintegration of an ice sheet, glacio-isostatic rebound of the crust in the area formerly covered by it is accompanied by the progressive collapse of the forebulge area.

proglacial outwash [19]: sediment accumulations deposited beyond the margins of glaciers and ice sheets by fluvial processes.

psu (Practical Salinity Unit) [77]: a unit measurement of ocean salinity based on the conductivity of seawater and expressed as grams per kilogram.

R

radar altimetry [7]: a method used to measure altitude of the ground surface beneath an aircraft or satellite, based on the transmission and return of radio waves.

radar interferometry [7]: sometimes referred to as Interferometric Synthetic Aperture Radar, is a technique of remote sensing used to measure surface elevation. It makes use of images of the Earth measured in different Earth orbits in the microwave spectrum, capable of penetrating through cloud masses, thus making it a valuable satellite technique to measure ground elevation.

radiocarbon dating [26]: the use of the radioactive isotope of carbon with a half-life of 5730 years to calculate the age of a sample. A calibration is needed to convert the ^{14}C age to a calendar age.

radiometric dating [27]: a process whereby the age of sediment is determined by the rate of decay of radioactive elements.

raised beach [10]: a collective term used to describe a range of relict coastal landforms located above present sea level.

raised shoreline [26]: sometimes used instead of the term, 'raised beach'. The term, 'raised shoreline' is also used in sea level change research as a theoretical construct to define an idealised or reconstructed position of former sea level across a given area.

relative marine transgression (or regression) [23]: an apparent rise (or fall) in the level of the sea in relation to the land.

relative sea level change [1]: an apparent change in the relative level of the sea that takes into account vertical changes in both the land and sea.

restrained glacio-isostatic rebound [58]: vertical rebound of the crust during a period of ice-sheet thinning that takes place while the landscape remains covered by the same ice sheet.

S

salinity of ocean water [8]: the amount of salt dissolved in ocean water and expressed in grams of dissolved salts per kilogram of ocean water. The salinity of ocean water varies regionally but has an average value of approximately 35 grams per kilogram of seawater.

saltmarsh [16]: a coastal ecosystem located across the upper part of the intertidal zone that is regularly inundated by tides.

salt weathering [11]: a process of mechanical weathering in coastal environments in which salt water penetrates into rock surfaces.

satellite altimetry [6]: a satellite method that uses radar pulses in order to determine the height of the ocean surface.

sea level curve [23]: a graph showing, for a particular location and time interval, an inferred pattern of relative sea level change.

sea level index point [23]: a sediment sample from a specific location of known age and altitude measured in relation to a datum level that can be used in the construction of a curve of relative sea level change.

sea level tendency [23]: a trend in a pattern of relative sea level change for a particular location towards, for example, increased marine or increased freshwater influence.

sediment bioturbation [45]: disturbance of sediment accumulations (in this case on the ocean floor) caused by the burrowing of organisms.

seismic discontinuity [31]: the boundary between the less dense crust with a seismic velocity of *c*.6kmsec^{-1} and the denser mantle, which has a seismic velocity in the

order of *c*.8kmsec^{-1} .

sensible heat [60]: the transfer of heat from one body to another that (unlike latent heat transfer) induces a change in temperature of the particular body.

shear rate [31]: the rate of change in the velocity at which one layer of a semi-fluid can move over an adjacent layer.

shear stress [31]: the force acting to induce deformation of a mass by slippage along a plane or planes parallel to the exerted stress direction.

shore platform [11]: a sub-horizontal coastal rock surface located between low and high water mark.

shoreline notch [10]: undercuts that typically form in carbonate rocks as a result of bioerosional processes.

sidereal years [29]: in sea level change research, sidereal years refer to calendar years that constitute the timescale against which other timescales (e.g. radiocarbon, Uranium–Thorium) can be calibrated.

solar insolation [20]: the amount of solar radiation that reaches the surface of the Earth.

sponge spicules [16]: structural elements characteristic of marine sponges and whose function is to provide protection against predators.

stable isotopes [21]: non-radioactive forms of atoms.

steric changes [8]: variations in sea level caused by short-term local (e.g. a storm) or longer-term global changes (e.g. thermal expansion of ocean water) in the temperature or salinity of a fixed mass of water in the world's oceans. Steric changes are associated with changes in ocean volume but no change in global ocean mass.

storm beach ridge [13]: a coastal landform generally composed of rounded and sub-rounded cobbles, shingle and sometimes sand. The ridges are typically produced as a result of storm wave action along coastlines characterised by long *fetch*.

strandflat [11]: broad coastal rock surfaces, typically several kilometres in width and characteristic features of many middle- and high-latitude regions including Norway, Scotland, Alaska the Antarctic peninsula and sub-Antarctic islands.

surface mass balance [61]: the difference between accumulation and ablation across a glacier or ice sheet. Surface mass balance changes are caused by changes in both temperature and snowfall. A glacier with a negative surface mass balance is in disequilibrium and will retreat, while a glacier with a positive mass balance is also out of equilibrium and will advance.

T

tectono-eustasy [33]: processes by which the volume of ocean basins is changed, mostly as a result of changes in the volume of ocean ridge systems and as a result of marine sedimentation.

tectonic uplift [36]: vertical movements of sections of lithosphere in areas of active plate motion.

testate amoebae [17]: a group of single-celled organisms known as protozoa that occur in a wide range of environments including saltmarshes.

theodolite [5]: a precision instrument used in surveying to measure horizontal and vertical angles.

thermal expansion of seawater [42]: the expansion of seawater due to changes in ocean water density which, in turn, is a function of changes in temperature and salinity.

thermohaline conveyor [81]: patterns of global ocean circulation driven by regional differences in surface ocean temperature and salinity. Surface and deep ocean thermohaline circulation patterns are connected by the production of Deep Water, the most significant of which in global terms are North Atlantic Deep Water (NADW) and Antarctic Deep Water (AADW).

tidal range [4]: the difference in elevation between high tide and the following low tide.

tide gauge [4]: an instrument, sometimes referred to as a mareograph, that measures elevation changes of the sea surface relative to a chosen vertical datum.

tidewater (outlet) glaciers [48]: glaciers that emanate from an ice sheet or ice cap and which reach the ocean. Such glaciers are commonly associated with areas of floating ice (ice shelves) and the production of icebergs.

trailing edge of a continental plate [41]: the edge of a continental plate characterised by low relief and an absence of seismic or volcanic activity.

transfer functions [16]: transfer functions enable quantitative estimates to be made of past changes in relative sea level, based on measured relationships between microfossil assemblages and tidal height.

tree ring chronology [29]: a timescale constructed on the basis of tree ring dating, also known as *dendrochronology*.

U

uranium–thorium dating [29]: a dating technique used to establish the ages of corals, speleothems, marls and tufa. The technique is based on the decay of uranium-234 (^{234}U) to thorium-230 (^{230}Th). The technique has an approximate age limit of c.350,000 years but is sometimes used to c.500,000 years through the use of thermal-ionisation mass spectrometry.

V, W, Y

viscosity [31]: a state in which a semi-fluid has an internal friction causing it to be 'sticky'. It is measured by the force exerted per unit area that resists a uniform flow.

WAIS [70]: West Antarctic Ice Sheet, a marine-based ice sheet, the bed of which lies well below sea level and whose margins merge into floating ice shelves.

wave-cut platforms [11]: coastal rock platforms produced as a result of wave abrasion, corrosion and hydraulic action.

Younger Dryas [56]: a period of exceptionally cold climate at the end of the Late Pleistocene and generally considered to have spanned the time interval between c.12.9 and 11.7kyr.

Selected further reading

Barth, M.C. and Titus, J.G. (1984) *Greenhouse Effect and Sea Level Rise: a Challenge for this Generation*. New York: Van Nostrand Reinhold.

Benn, D.I. and Evans, D.J.A. (2010) *Glaciers and Glaciation*. London: Hodder Education (Chapter 6 'Greenland and Antarctic Ice Sheets' and Chapter 7 'Glaciers and Sea-Level Change' are particularly informative).

Church, J.A. and Woodworth, P.L. (2010) *Understanding Sea-Level Rise and Variability*. Hoboken, New Jersey: Wiley-Blackwell.

Dawson, A.G. (1992) *Ice Age Earth: Late Quaternary Geology and Climate*. London: Routledge.

Devoy, R.J.N. (ed.) (1987) *Sea Surface Studies*. London: Croom Helm (a classic of its time covering all key topics).

Hay, W.W. (2016) *Experimenting on a Small Planet: a history of scientific discoveries, a future of climate change and global warming*. New York: Springer (a fascinating read with a very useful chapter 27 on 'Sea Level').

Lowe, J.J. and Walker, M.J.C. (2014) *Reconstructing Quaternary Environments*. (3rd edition) London: Routledge (provides useful overviews of dating methods in Chapter 5 and sea level change in Chapter 2, section 2.5).

Murray-Wallace, C.V. and Woodroffe, C.D. (2014) *Quaternary Sea Level Changes*. Cambridge University Press.

Pirazzoli, P.A. (1996) *Sea Level Changes: the Last 20,000 years*. Hoboken, New Jersey: Wiley.

Sissons, J.B. (1967) *The Evolution of Scotland's Scenery*. Edinburgh: Oliver and Boyd (classic text with informative account of processes of relative sea level change in an area affected by glacio-isostatic rebound).

Smith, D.E. and Dawson, A.G. (1983) Shorelines and Isostasy. London: Academic Press.

Stow, D. (2017) *Oceans: a Very Short Introduction*. Oxford University Press.

Douglas, B.C., Kearney, M.S. and Leatherman, S.P. (2000) *Sea Level Rise: History and Consequences*. London: Academic Press.

Useful websites

NASA Sea Level Change Portal: https://sealevel.nasa.gov

Watts Up With That? Inconvenient: NASA shows global sea level… pausing, instead of rising: https://wattsupwiththat.com 2017/10/16 (an alternative view).

Antarctic Glaciers: http://www.antarcticglaciers.org (contains several useful sections that discuss glacier fluctuations and sea level change).

ARGO ocean observations portal: http://www.argo.ucsd.edu (global monitoring of ocean salinity and temperature for different ocean depths – data freely available to public).

University of Colorado (CU) Sea Level Research Group: http://www.sealevel.colorado.eu (an excellent informative site).

Intergovernmental Panel on Climate Change (IPCC) http://www.ipcc.ch Click on 'Publications and Data', then download the respective 'Physical Science Basis' Working Group Report from the 5 Assessment Reports that cover the period 1991-2013 and search the appropriate 'sea level' sections.

The Mitrovica Group: http://mitrovica.eps.harvard.edu (an excellent starting point for those wishing to learn more about glacio-isostatic adjustment and sea level change).

University of Durham Sea Level Research Unit (SLRU): https://www.dur.ac.uk (a good place to learn about different methods of sea level change reconstruction).

Penn State University - Absolute Versus Relative Sea Level Change: https://www.e-education.psu.edu (contains a wide range of fascinating material on sea level change science).

National Snow and Ice Data Center – Contribution of the Cryosphere to Changes in Sea Level https://nsidc.org (presents a different perspective on contribution to sea level rise of ice in cryosphere).

National Ocean Service (NOAA) – https://oceanservice.noaa.gov (useful discussion of role of thermal expansion of seawater in sea level rise).